AMAZING FACTS – AUSTRALIA'S SOUTHERN SKIES

CONTENTS

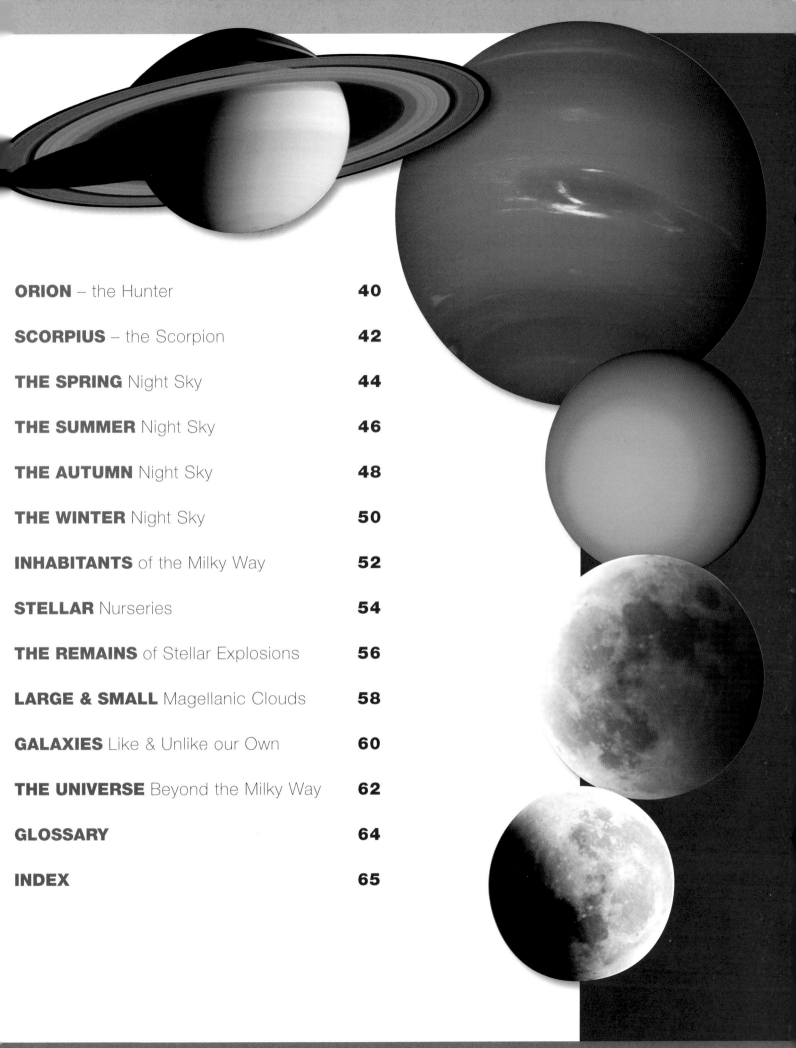

MAGAZINES – These will not only keep you up to date on new discoveries in astronomy, they will also provide information on happenings in the night sky. In Australia, the premiere magazine on astronomy is: "Sky & Space" *(www.skyandspace.com.au)*. Another excellent choice is "Sky & Telescope" *(www. skypub.com)*.

ASTRONOMY CLUBS AND ORGANISATIONS – A great way to learn about the night sky is to spend time in the company of other amateur astronomers. Most cities have at least one such group. An excellent starting place is the Australian Astronomy website *(www.astronomy. org.au)* which links to amateur astronomy societies all over Australia. Try to attend a "star party" – they are frequently scheduled on weekend nights around the New Moon when the skies are darkest.

LECTURES – Universities with astronomy programs run public lectures given by faculty or visiting professors.

OBSERVATORIES AND PLANETARIA – All planetaria and many observatories have public programs. A list is maintained on the Australian Astronomy website *(www. astronomy.org.au)* as well as by Quasar Publishing *(www. quasarastronomy.com.au)*.

YEARLY GUIDES
– Quasar Publishing *(www. quasarastronomy.com.au)* produces a yearbook of astronomy.

The 8 m Gemini South telescope, in which Australia is a partner, stands ready to begin a night of observing the skies from the top of Cerro Pachon in Chile.

INTRODUCTION
to Astronomy

The Universe is endlessly fascinating – there is so much that is already understood, but even more new things to discover.

Every amateur astronomer dreams of the day when they can finally view the southern sky, because the southern sky really is more exciting than the sky that is visible in the northern hemisphere! Many of the best and brightest objects of different kinds can be best seen, or only seen, from the south. For you to fully appreciate the southern sky, we must introduce some new words and concepts that will help you to understand astronomy. Many of the scientific words used in the text are further explained in the glossary at the end of this book.

Scientists refer to anything composed of atoms as matter. Matter may be found in

one of three different states: solid, liquid, and gaseous. This book is an example of solid matter, the ocean is liquid matter, and the air around us is gaseous matter.

The atmosphere is the entire region of gaseous matter (gas) that is found above the surface of a planet. The gases hydrogen and helium account for 99% of the atoms in the Universe, even though they are not the most common atoms on Earth. The vast black area between stars, which we call space, is very empty – it contains less than one atom per cubic metre.

Gravity is the property of matter in the Universe that causes it to be attracted to other matter. In our solar system, the Sun contains most of the matter, so the motions of the planets, whose atoms are affected by the Sun and subject to gravity, are altered into closed loops, called orbits.

Apart from regular matter, energy exists in our Universe in another form – a form called light. Information from the skies arrives on Earth as light. Light comes in many colours, which correspond to different energies. Regardless of its colour, light travels very very rapidly – 300,000 km every second!

The distances between objects in the sky are so large that we usually measure distance by how long light takes to travel from one place to another. A person standing 3 m away from you is $1/100,000,000$th of a light-second away. Australia is just over $1/100$th of a light-second across. The Moon is just over one light-second away. The most distant planet visible to the unaided eye is Saturn, at just over one "light-hour" away. The most distant planet in our solar system, Pluto, is four light-hours away.

Outside of the planets orbiting our Sun, it is a long way to alpha Centauri, the next nearest star, which is 4.3 "light-years" away. Both our Sun and alpha Centauri are part of a much larger group of about a million, million stars that are known as our Galaxy. The shape of our Galaxy is a disk with a roundish bulge of stars near its centre. In the night sky, we see the more distant stars in our Galaxy as a faint band of light called the Milky Way. The diameter of our Galaxy is roughly 100,000 light-years.

The Universe contains billions and billions of galaxies, which come in all sorts of sizes. Two much smaller galaxies orbit the Milky Way, much like the planets orbit the Sun; they are known as the Large and Small Magellanic Clouds and are 170,000 and 200,000 light-years away. The nearest galaxy like our own is 2.2 million light-years away. The most distant galaxies observed by telescopes to date are about 13 billion light-years away!

The Universe is a HUGE place! Humans have so far only reached distances of one light-second through manned space exploration. To discover more, astronomers must rely on what can be learnt from light when studying the stars and space.

SCIENCE – Astronomy is one of the few areas of modern science where an amateur can make a useful contribution. Observations of variable stars can be made visually or with imaging equipment. The following organisations have active variable star programs: Astronomical Society of South Australia *(www.assa.org.au/ sig/variables);* Royal Astronomical Society of New Zealand *(www.rasnz.org. nz);* American Association of Variable Star Observers *(www.aavso.org).*

INTERNET – There are thousands of excellent astronomy-related websites on the internet. Excellent starting points are magazine, observatory, university, or space agency sites. Spectacular images are available for free download and some observatories even have "virtual tours".

The northern hemisphere sky is seen spinning around the north celestial pole above the 8 m Gemini North telescope's dome on Mauna Kea in Hawaii. Australia's participation in the Gemini Observatory gives its astronomers large-telescope access to the objects in both hemispheres.

STARING INTO SPACE

TELESCOPES AND BINOCULARS

The southern skies can be explored and enjoyed using binoculars. In fact, many objects described in this book are best seen with binoculars. When choosing binoculars, look for a description of about "7 x 50". The first number is the magnification and the second is the diameter of the front lens in millimetres. Any binoculars can be used for astronomy, but the best binoculars will be those that gather the most light, which is determined by the diameter number. Low magnification usually means a wide field and the lens size is large but not too heavy. A higher magnification power will also magnify any unsteadiness of your hands, so don't choose a power higher than ten. Remember, you will be lifting binoculars over your head to view the southern skies, so choose a weight that you can handle comfortably.

There is lots of equipment for amateur astronomers available through specialty and camera stores. Advertisements can be found in "Sky & Space". The recommendations of other amateur astronomers should also help. Binoculars are an excellent first instrument. It is now possible for amateurs to buy computer-controlled telescopes for $1000–3000! But it is usually possible to find "used" telescopes advertised at a good discount from their retail prices.

The author, Doug Welch, at age 15 with his 20-cm reflecting telescope at twilight.

THE NEAREST STAR
– the Sun

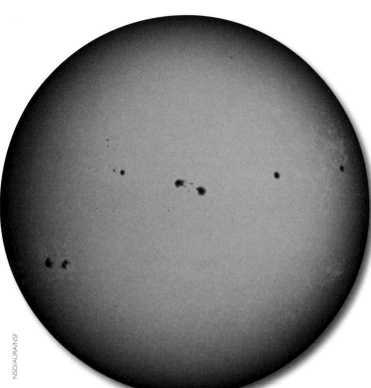

The Sun is a round, hot ball of gas and is too bright to look at without special equipment. Its energy fuels life on Earth.

NSO/AURA/NSF

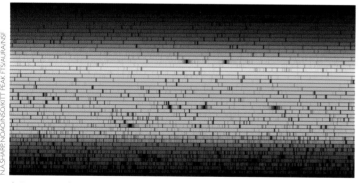

N.A.SHARP, NOAO/NSO/KITT PEAK FTS/AURA/NSF

Astronomers learn about the Sun by studying its light. They figure out how hot it is and what it is made of by studying the missing colours in sunlight.

It's hard to believe that the huge, bright hot Sun in the daytime sky is the same as the pinpricks of light that we call stars in the night sky. But it is true: the Sun is just another star!

THE DIFFERENCE in appearance is simply because of how very much closer we are to our particular star, the Sun.

The next closest star, alpha Centauri or Rigil Kent, is visible only in the southern sky – later in this book you will find out where alpha Centauri is located. To understand how incredibly distant even this closest star is, pretend that the distance of the Earth from the Sun is just 1 m. By this measure, alpha Centauri would still be 270 km away! If you picked a random star in the sky (not one of the brightest or faintest you could see with your eyes alone) its distance would be a few hundred to 2000 light-years away. Much further indeed than alpha Centauri's 4.3 light-years.

We feel no warmth from fainter stars because of their distance from the Earth. As you get farther away from a source of light, your eyes see a smaller and smaller fraction of the light given off, and you feel less heat. We receive 10,000,000,000 times more light from the Sun than we do from alpha Centauri because of the difference in distance!

HOW BIG IS THE SUN?

We can see that stars like the Sun are round, but just how big is it? The Sun is very, very big compared to the Earth. If you imagine the Earth as being the size of a pea, the Sun would be a ball of 5 m across by comparison.

WHAT IS A STAR MADE OF?

THE PROPERTIES OF A STAR are determined by one thing: its mass, or how much matter it contains. Stars are made entirely of gas. The mix of atoms is surprisingly similar from star to star. Almost all stars are made of 90% hydrogen, 9% helium and the remaining 1% of carbon, nitrogen, and oxygen with a smattering of many other elements. Stars are different from planets because they have enough mass for nuclear reactions to occur at their centres. This enables them to produce their own energy, and therefore their own light. Planets are only visible when they reflect a star's light. The amount of light a star gives off depends on its mass. The most massive stars have 100 times the mass of the Sun and the least massive are 0.08 of the Sun's mass. Low-mass stars are by far the most common. Stars with a large mass give off WAY more light than low-mass stars. This usually means that massive stars, even those at a great distance, are easier to see than closer low-mass stars. Stars don't have a solid surface, but they have a well-defined size. Massive stars have outer layers of gas at a temperature of 100,000°C. The lowest-mass stars are only about 2000°C on the outside.

WHY IS THE SUN SO DIFFERENT?

The Sun is different from the Earth in several ways. First, it is incredibly hot on the outside – close to 6000°C – and even hotter inside. Second, it has no solid surface. Third, unlike the Earth, the Sun produces its own energy deep down in its centre. Due to the high temperature affecting the atoms that make up the Sun, they collide at high speeds, react and release energy – a process known as fusion. Only objects with the size and makeup of stars can produce energy by fusion and all starlight is created in this way. The Sun is a very active place. Darker regions on the Sun are called sunspots, and change from hour to hour. They are the equivalent of storms on the Sun. Sometimes these sunspot storms create large explosions which are so violent that they send a tiny fraction of the Sun's material off into space. When this material reaches the Earth it can result in displays of "Aurora Australis" – the southern lights.

HOW BRIGHT IS THE SUN?

It may seem that the sun changes in brightness every day, being brighter when it is high in the sky and fainter near sunrise and sunset. But the Sun is not changing, it is the amount of air through which the sunlight passes in order to reach the Earth that changes.

The Sun appears much redder when it is close to sunset, but this is also an effect of the Earth's air. Sunlight contains all of the colours of the rainbow, but red light filters through the Earth's air more easily than blue light, which makes sunsets appear red.

THE SUN'S EFFECT ON LIFE

Life on Earth exists because of energy projected from the Sun in the form of light. This energy heats the ground, the oceans and the atmosphere and triggers many chemical reactions, such as the production of oxygen by green plants.

Without the Sun's energy, the Earth would become a lifeless planet: all of the oceans and liquid water would freeze and heat from the Earth would flow out into space.

Fortunately for life forms on Earth, the Sun is a very reliable source of energy and is expected to continue producing energy for another five billion years.

As we talk about the many objects that orbit the Sun, keep in mind that it is only because of the light and heat given off by the Sun that we exist to see the planets and constellations at all!

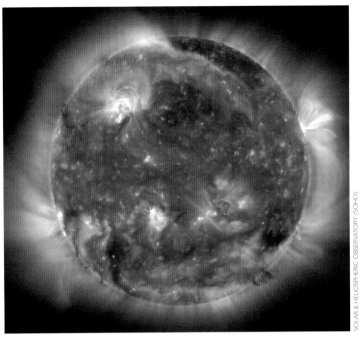

Gas streams off the surface of the Sun, as seen in this image taken from space.

SOLAR & HELIOSPHERIC OBSERVATORY (SOHO)

How to view the Sun – Never stare at the Sun, either with your own eyes or through binoculars or a telescope. Doing so will damage your eyes and can leave a permanent dead spot in your vision. Astronomers study the Sun using special instruments.

MIKE GAYLARD, HARTRAO

Energy from the Sun in the form of light stops the oceans from freezing and provides the chemical energy for life.

DOUG WELCH

THE MOON is 385,000 km away from the Earth and orbits it at 3600 km/h.

UNLIKE THE EARTH, the Moon has no atmosphere, so there is nothing to stop the rocks that orbit the Sun from smashing directly into the Moon's surface.

DEPENDING ON the Moon's circuit around the Earth, it can sometimes be seen in the daytime or night-time sky.

THE MOON APPEARS very bright against the dark night sky but its surface is actually a very dark grey.

THERE IS NO permanent "Dark Side of the Moon". All parts receive light half of the time during each orbit.

There is a **near** and a **far** side, but because the moon rotates once each orbit, we always see the near side.

FRICTION FROM the ocean tides on Earth slowly lengthens the day on Earth, so the size of the Moon's orbit slowly increases. Each year the average distance to the Moon increases by 3–4 cm. We are able to measure the Moon's distance to this precision!

The Earth and the Moon seen at a great distance by Mariner 10 on its way to Venus and Mercury. Note how much smaller the Moon is than the Earth.

Binoculars or a telescope allow you to see the Moon in much greater detail, as in the picture above on the left. Without magnification, only the largest dark and light features are seen, as in the smallest image above.

THE Moon

The Moon doesn't actually give off its own light – it simply reflects light from the Sun.

WE SEE THE MOON only because of the Sun, and all of the planets and objects in our solar system are visible only in this way.

The Moon may appear to be around the same size as the Sun, but that is only because the Moon is much closer to the Earth. The Moon is really a lot smaller – just under 3500 km in diameter. The Sun is 375 times farther away than the Moon, so the Sun must also be 375 times larger!

It takes the Moon 28 days to orbit the Earth once, which is the same amount of time it takes for the Moon itself to spin once around. As a result, we always see the same half of the Moon and never the other side.

When viewing the Moon with the unaided eye, you will see darker and lighter patches that are sometimes called the "Man in the Moon", because they appear to be round eye shapes above an open mouth. These darker parts are actually large, smooth areas of cooled lava that resulted from large rocks smashing into the Moon at high speed billions of years ago.

With a pair of binoculars or a small telescope, you can see much more detail, including mountains and craters. The best time to see detail with these instruments is at First Quarter, when the shadows cast by the crater walls and other features give the Moon a stunning three-dimensional look.

The Moon is smaller and less dense than the Earth and, as a result, has just 1/6th the gravity. So you could jump six times higher and six times farther on the Moon (if you didn't have to wear a spacesuit)!

On Earth, temperatures differ by only 10–20°C at night. This is because during the day our atmosphere traps some of the Sun's heat then releases it slowly at night. There is no atmosphere on the Moon, so the temperature difference between sunlight and shade is dramatic – over 200°C!

The craters on the Moon are mostly round and most were formed billions of years ago. The few that formed more recently are whiter and brighter. Scientists thought the older craters had been formed by volcanoes, like some of the craters on Earth. It was mistakenly thought that if the craters were due to the impact of rocks from space, some should be elongated. We now know that the speed of the impact would have been so great – at 40,000–160,000 km/h – that the incoming rock would have created an explosion below the surface.

MOON PHASES

Earth's gravity is strong enough to keep the Moon circling the Earth at a distance of 385,000 km away. As the Moon orbits the Earth, sunlight arrives on the Moon in a different direction from our viewpoint from Earth, and as the Moon and the Earth are both in orbit, that direction changes from day to day.

Like everything else in the solar system, the half of the Moon facing the Sun will be illuminated and the other half will be seen as dark.

When the Moon is in the same direction in the sky as the Sun, we cannot easily see it – this phase is called "New Moon".

A few evenings later, a thin, bright crescent will be visible. When the Moon appears to be half-illuminated, we call it "First Quarter" because it is one-quarter of the way through its circuit around the Earth.

A "Full Moon" is when the Moon is opposite the Sun in the sky and it will rise at sunset and set at sunrise.
A half-lit Moon in the morning sky is called "Last Quarter". About seven days later, the circuit is complete and it is a "New Moon" again.

The Davy crater chain photographed by Apollo 12 astronauts orbiting the moon. Crater chains occur when an already fragmented object hits the surface.

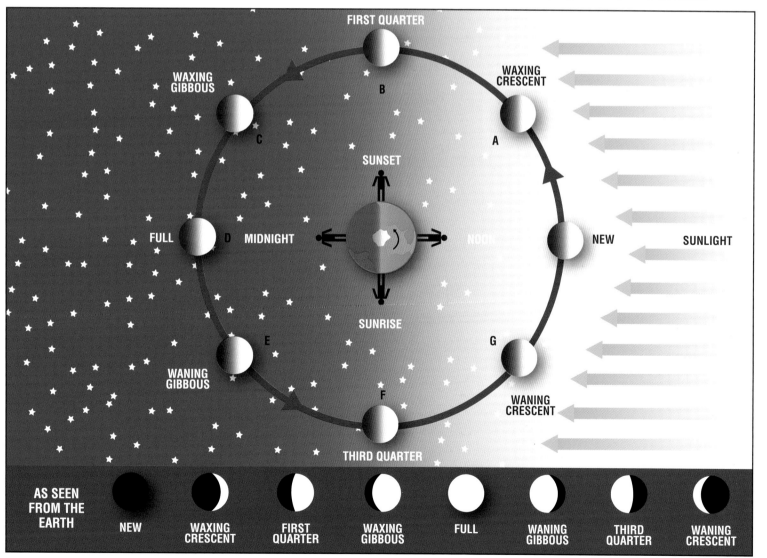

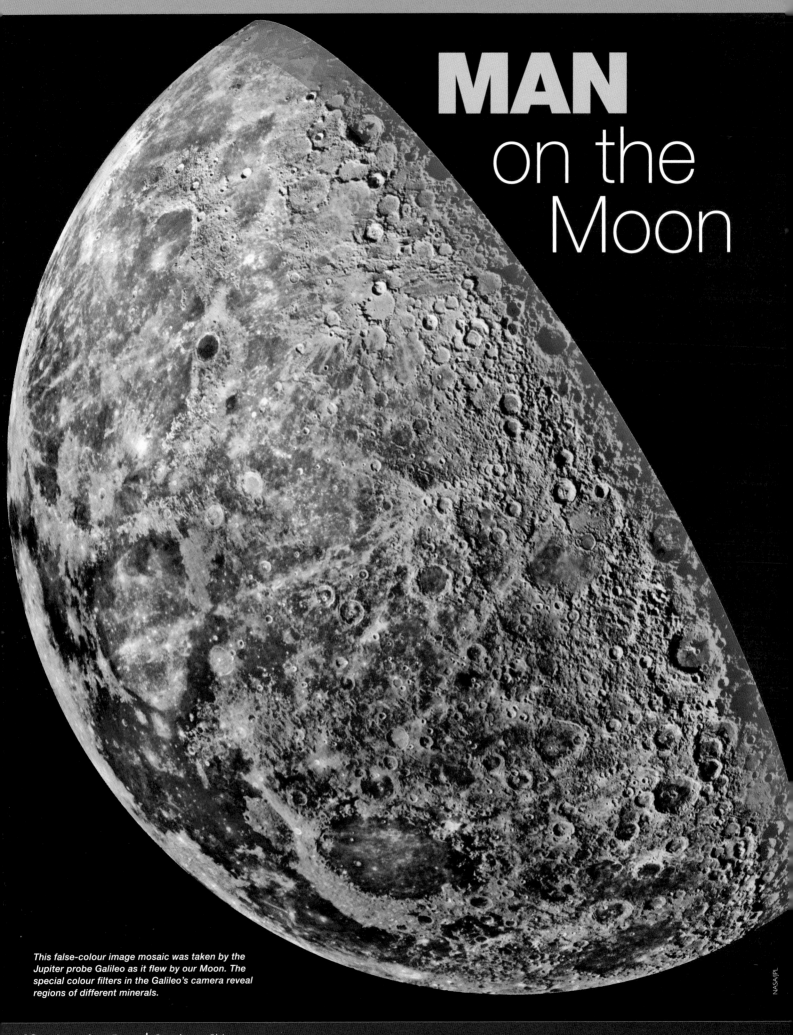

MAN
on the
Moon

This false-colour image mosaic was taken by the Jupiter probe Galileo as it flew by our Moon. The special colour filters in the Galileo's camera reveal regions of different minerals.

NASA/JPL

MANNED VISITS TO THE MOON

Between 1969 and 1972 there were six manned-landings on the Moon by the United States. Twelve astronauts have walked on its surface, deployed equipment, collected rock samples and returned to Earth. The Soviet Union, as it was known, explored the moon with robotic landers, also returning moon rocks to Earth. These expeditions taught us a great deal about the Moon's age and history. We learned that the mountainous parts of the Moon are the oldest (between 4.0 and 4.5 billion years old) and the large, smooth lava-filled basins called "maria" are the youngest (3.5 billion years old). Most of the cratering occurred before the maria were formed.

The bootprint of Apollo 11 astronaut Buzz Aldrin on the lunar surface. Note how fine the lunar soil is and how easily it cakes.

RETURNING TO THE MOON

It has been over 30 years since a person last walked on the Moon. Several countries are planning new missions, some manned, some using robots. One reason for the renewed interest in lunar exploration is recent evidence that water ice may have collected in the permanently shaded craters near the poles of the Moon. Water is required for any manned mission to the Moon or to any other object in the solar system. In the future, water and raw materials for missions to other planets may be extracted from the Moon, rather than transported from the Earth, because launching supplies from the weaker lunar gravity requires far less rocket fuel than launching from Earth. In the past, all manned missions had to take everything they needed with them and consequently could not stay for long before having to return to Earth.

Astronaut Buzz Aldrin poses on the Moon during the Apollo 12 visit.

MOON ROCKS

Imagine the thrill of holding a piece of the Moon in your hand! What would it be like? Surprisingly, the minerals found in rocks on the Moon are also found here on Earth. The rocks on our home planet, however, contain many minerals that aren't found on the Moon.

One main difference is that molecules of water aren't found in lunar minerals. On Earth, rocks are usually classified as either igneous, sedimentary, or metamorphic. On the Moon all rocks are igneous. Apart from the lack of water molecules, the material making up the Moon is very much like the crust of the Earth. Many astronomers currently believe that the Moon was once the crust of the Earth until a huge impact tore off material from the outer layers of the Earth which then collected together to form the present Moon.

Apollo 17 astronaut Eugene Cernan checking out the Lunar Rover that enabled them to explore the moon.

Images of Earthrise, as seen by the Apollo 8 astronauts orbiting the moon, touched human emotions.

DID YOU KNOW?

OVER THE COURSE of millions of years, Mercury's orbit changes from nearly perfectly circular to twice as elliptical as it is now.

A TIN CAN placed on the sunlit side of Mercury or the surface of Venus would melt.

A PROBE named Messenger is on its way to orbit Mercury in 2011. Only one spacecraft, Mariner 10, has visited Mercury before. It flew past it three times during 1974 and 1975. To go into orbit around Mercury, Messenger needs to alter its own orbit to closely match the planet's. Instead of using rocket fuel to achieve this, Messenger will make close passes to Venus (twice) and the Earth (once) to alter its orbit using "gravitational assists" from those planets.

ON THE SURFACE of Mercury is a large impact feature called the Caloris Basin. The largest ring of this basin is 1300 km across. An impact this big would have probably significantly changed much of the landscape of Mercury.

IT TAKES 116 DAYS for Mercury to be seen in the same position relative to the Sun on Earth. If Mercury is visible in the evening on a given date, it will appear again in the evening just under four months later.

MERCURY REFLECTS only about 11% of the light that hits it, making it one of the darker surfaces in the solar system, comparable to the Moon.

MERCURY has a magnetic field, but it is only 1% of the strength of the Earth's.

MERCURY & VENUS
the Twilight Planets

The two planets closest to the Sun never appear very far from it in the sky, so they are visible only in the few hours before sunrise or after sunset.

The huge impact feature on Mercury known as the Caloris Basin is half-visible near the shadow boundary at the centre left of this image, obtained by Mariner 10.

NASA/JPL

MERCURY ORBITS the Sun with an average speed of 173,000 km/h, but the planet itself spins very slowly – so slowly that a day on Mercury (from sunset to sunset) takes two Mercury years (two orbits of the Sun). The daytime temperature on Mercury's surface is more than 400°C; at night it is colder than -180°C.

With a diameter of 4900 km, Mercury is larger than the Moon but smaller than the Earth. It is much denser than the Earth and has a large central core that is primarily made up of the minerals iron and nickel.

TO SEE Mercury it helps to know that the planet is at the most visible place in its orbit when the sky is darkest. Venus, however, is the brightest object in the sky besides the Sun and the Moon, making it very easy to see. Because it orbits the Sun at a greater distance than Mercury, Venus also appears in the darkened twilight sky for longer.

PLANET MERCURY

Mariner 10 gave us our first images of Mercury's surface, which looks very much like that of the Moon. There is no atmosphere to protect the planet, so it is pockmarked with many craters. Being close to the Sun, the pull of the Sun's gravity is stronger on Mercury, meaning that objects which strike the surface deliver more energy than those that collide with the Moon.

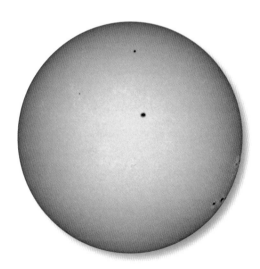

NSO/AURA/NSF

On May 7, 2003, Mercury crossed the Sun's disk, an event seen from the Earth. It is the small, circular black dot in the top centre of the Sun in this image. Mercury never appears larger than this from Earth and since it is always close to the Sun, no satellite telescope can observe it. Irregular dark patches on the sun are sunspots.

This Australian sunset image reveals (upper left to lower right) Venus, the crescent Moon, Jupiter and Mercury.

PLANET VENUS

The second-closest planet to the Sun resembles the Earth in size and total mass, but little else! For many decades, astronomers could only guess what the surface might be like because the entire planet is covered in dense cloud. Using new detectors, it was found that Venus is very hot – far hotter than could be accounted for by simply being nearer to the Sun than the Earth! We now know that Venus is so hot because gas in the planet's atmosphere, mainly carbon dioxide, stops energy escaping back into space. The surface temperature is 480°C – hotter than the hottest spot on Mercury and hot enough to melt lead! Even stranger is the fact that the night-time and daytime sides of Venus have nearly identical temperatures.

Volcanoes and only large craters are found on the surface. Small craters aren't formed because only the largest rocks survive the trip through the thick atmosphere. The surface of the planet is also very young since the volcanoes continually replenish it.

The Magellan probe orbited Venus and measured the altitude of surface features using radar. This reconstructed hemisphere of surface detail is impossible to view directly due to thick clouds.

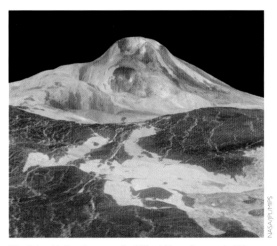

The 8 km high volcano called Maat Mons is seen in this image reconstructed from radar measurements made by the Magellan probe. The viewpoint is 560 km north and 1.7 km above the Venusian terrain.

Three impact craters, 40–60 km in diameter are seen on the Venusian surface in this three-dimensional reconstruction from radar data taken by the Magellan probe.

THE FACTS!

THE ATMOSPHERE of Venus holds 100 times as much gas as the Earth's atmosphere, but none of it is oxygen.

VENUS spins very slowly in the opposite direction to the Earth.

THE HIGH TEMPERATURES on the surface of Venus are due to its carbon dioxide atmosphere trapping energy from the Sun. This is called the "runaway greenhouse effect".

THE SURFACE of Venus has been mapped using radar. The earliest radar maps were made using radar pulses sent from the radio telescopes on the Earth.

IF YOU KNOW just where to look, it is easy to spot Venus during the daytime. Binoculars make it even easier.

THE BRIGHTNESS of Venus in the night sky is due to its white cloudtops. They reflect three-quarters of the light hitting them.

THE CLOUDS of Venus are made up of sulphuric acid droplets, not water.

IT TAKES VENUS 584 days to return to the same position in the sky relative to the Sun (as seen from Earth). This period served as the basis for the religious calendar of the Central American Mayan civilisation.

THE ORBIT OF VENUS is very close to circular, so the apparent size of the Sun changes very little.

VENUS and Mercury are the only planets in the solar system that do not have moons.

Above: A close-up of the moon Phobos, one of two small moons orbiting Mars, captured from less than 200 km away by the Mars Express spacecraft.

Below: The Hubble Space Telescope orbiting Earth took this image during its very close approach of Mars in 2003.

NASA, J. BELL (CORNELL U.) & M. WOLFF (SSI)

Below: This image captured by ESA's Mars Express orbiter apparently shows crater floor material flowing between two craters – indicating the presence of ice.

MARS
the Red Planet

When Mars is visible in the evening sky, there is no mistaking it. The planet appears a bright red-orange colour and moves noticeably from night to night against the backdrop of the distant stars.

EARLY ASTRONOMERS fuelled the public's imagination about what might be on Mars. It was known from studies with telescopes on Earth that a day on Mars lasts 24.6 hours – very similar to one Earth day. Also, Mars clearly had polar icecaps and coloured surface features that change with the Martian seasons. Many people believed that life, possibly even advanced life, might be present on Mars.

Analysis of the light from Mars by ground-based telescopes long ago ruled out Mars being very much like Earth. It was found that the amount of water vapour in the atmosphere of Mars was very low and that the atmosphere had only about $1/100$th of the gas found in the Earth's atmosphere. Oxygen was also virtually absent in the Martian "air".

All of these findings suggested that advanced life was practically impossible at the present time on the Red Planet.

The water that exists on Mars isn't in liquid form on the surface because the atmospheric pressure and temperature on Mars allow only the gaseous form (water vapour) or the solid form (water ice). It is now known that the polar caps of Mars do contain large quantities of water ice, but that so-called "dry ice" (solid carbon dioxide) forms on the polar cap during the Martian winter.

Space probes that arrived on Mars found a fascinating world of enormous volcanoes, giant canyons, craters, and planet-wide dust storms – but one without running water or any obvious sign of life. Strangely, there *is* evidence of erosion caused by running water, which seems contradictory until you allow for the possibility of underground water ice melting due to a source of sudden heating. The walls of valleys and craters on Mars suggest that water ice is very common near the Martian surface.

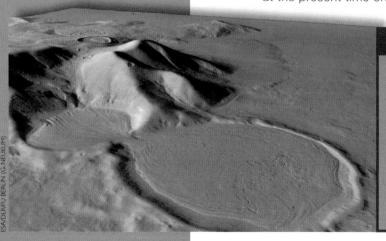

WHAT IS MARS LIKE ON THE INSIDE?

The crust (the outermost solid layer) of Mars is several times thicker than the Earth's crust. It covers a liquid core that occupies about one-quarter of the planet. We understand the interiors of the Earth and Moon better because we have been able to use seismometers located at several different places to measure how sound is transmitted through those bodies. Mars has resisted study with this technique to date – both Viking 1 and 2 had seismometers on them, but only the one in Viking 2 functioned. It recorded a single "Marsquake" before being shut down.

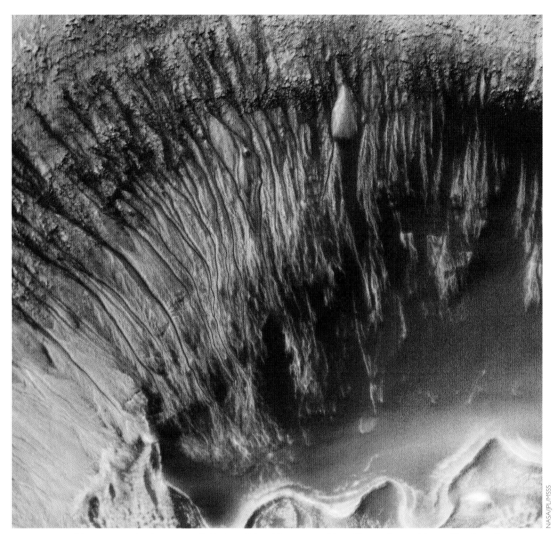

NASA/JPL/MSSS

This crater wall image was obtained by NASA's Mars Global Surveyor orbiter. The gullies and finger-like deposits suggest that flows of liquid water and debris were involved. Water would only exist briefly in such situations, evaporating or freezing very quickly.

MARS HAS only 11% the mass of the Earth and only 15% of the volume.

THE SURFACE gravity of Mars is 38% of the Earth's. You could jump 2.5 times higher on Mars than on Earth – if you didn't need a spacesuit!

MARS ORBITS the Sun at a distance of 228 million km, on average. The orbit of Mars is less circular than the Earth's so there is a greater range in how much energy it receives from the Sun during a Martian year. At its closest, Mars is 207 million km from the Sun. At its most distant, it is 249 million km away.

MARTIAN DUST STORMS can engulf the entire planet. The dust must be exceedingly fine to stay afloat in the very thin Martian atmosphere.

THE ATMOSPHERE of Mars is 95.3% carbon dioxide, 2.7% nitrogen, 1.6% argon and only 0.13% oxygen.

Below: A true-colour close-up of a rock on the surface of Mars, obtained by NASA's Mars Pathfinder lander.

THE MARS WE SEE TODAY may be quite different from the Mars of hundreds of millions, or even billions, of years ago. The gravity of Mars is so weak that most of the planet's atmosphere will escape given time, but there may have been periods in ancient Martian history when gas was supplied from surface activity (for example, by volcanoes) at a faster rate than it was being lost to space. If the atmosphere was thick enough, at some point liquid water could have existed. Under those conditions, early forms of life might well have developed.

AN IMPORTANT QUESTION that we may be able to answer by sending landers to Mars is: how easy is it for simple life forms to form? We know that life on Earth has built up over billions of years by the joint forces of chemistry and natural selection. Clearly, conditions on Mars haven't remained favourable for life, but it is possible that the most basic forms of life once existed on Mars – by studying surface and sub-surface samples from the planet we can find out whether or not that happened. If it did, then the implications are huge for the abundance of life elsewhere in the Milky Way and the rest of the Universe. It would likely mean that the development of life is inevitable and that on planets that have long-lived atmospheres, the existence of higher life forms is actually fairly common!

NASA/JPL

DID YOU KNOW?

THE CHANGE in season is much more pronounced on Mars, due to its orbit being elliptical and the planet being tilted by 25.2 degrees to its orbit (which is greater than the 23.5 degrees tilt of the Earth).

THE AVERAGE temperature of the Martian surface is -63°C, but over the course of one Martian day, it can range from -89° to -31° C because the thin atmosphere of Mars does not hold heat well.

ON AVERAGE, a hectare on the surface of Mars will receive only 43% of the energy from the Sun that the same area receives on Earth.

A DAY on Mars is 24.6 hours long. The fact that this is so similar to the length of the Earth's day is simply a coincidence.

Below: This image by Viking 1 Orbiter reveals how large the rift valley called Valles Marineris is with respect to the whole planet. At 4000 km long and 10 km deep, it is the size of a continent on Earth.

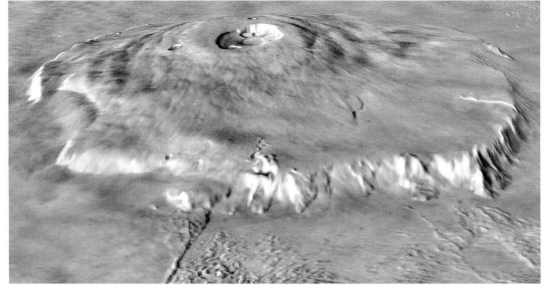

The largest volcano on Mars, Olympus Mons, is shown is this specially processed image where height is exaggerated by ten relative to the other dimensions. Olympus Mons is the largest volcanic feature on any planet in the solar system.

MARS IS ORBITED BY TWO MOONS, but they are very different from Earth's Moon. Phobos and Deimos, as they are known, are only 50 and 100 km across, compared to our Moon's 3500-km diameter. They aren't even round! Both are believed to have formed elsewhere in the solar system and to have been captured by the planet's gravity. Phobos orbits Mars in just 7.7 hours – far shorter than the Martian day – which has the peculiar effect of causing it to rise in the west and set in the east. Deimos is farther away and orbits Mars in 30.4 hours. They both orbit the planet at distances far smaller than the Earth–Moon distance and their orbits are inclined very little to the orbit of Mars, which results in very frequent solar and lunar eclipses on Mars. However, neither moon appears large enough in the sky to block out the entire disk of the Sun as seen from the surface of the planet.

MARS ALSO HAS a number of large volcanoes. The largest of these, Olympus Mons, is also the largest volcano in the solar system. It is 24 km high, nearly three times

NASA/USGS

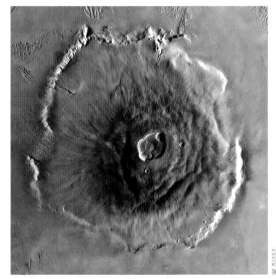

Olympus Mons viewed from above by the Viking 1 Orbiter.

the height of Mount Everest, the highest peak on Earth. Olympus Mons is so tall that the summit towers above most of the Martian atmosphere! Much to the surprise of scientists studying Mars, recent evidence suggests that Olympus Mons has been active for about 80% of the time since the planet formed. It also appears that it was last active only a few million years ago.

VALLES MARINERIS is a spectacular canyon spanning a full 4000 km on the Martian surface. There are places where it is as deep as 10 km. However, Valles Marineris was not carved by water erosion like well-known canyons on the Earth. Early in Martian history there was an enormous accumulation

16 Amazing Facts | Southern Skies

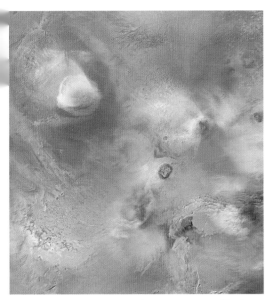

Images taken by Mars Global Surveyor during a Martian day show volcanoes (partially hidden by water-ice clouds) to the left and the Valles Marineris to the right.

ROBOTIC VISITORS TO MARS

Images of Mars have been sent back by many space probes, but other missions have not been so lucky. The failure rate of Mars missions is more than 50%! The first seven missions by the former Soviet Union failed by not reaching Earth orbit, only reaching Earth orbit, having radio failures, or landing during a Mars dust storm. The first US mission, by Mariner 3 in 1964, also failed, but Mariner 4 successfully flew past the Red Planet and sent back 21 images. The first successful landers were NASA's Viking 1 and 2, in 1976. Since then, three lander/rovers have returned images and data: NASA's Mars Pathfinder in 1996 and Mars Exploration Rovers Spirit and Opportunity in 2004.

THE FACTS!

IT TAKES Mars 687 Earth days to orbit the Sun once. The fact that the Earth is moving along its smaller orbit at a greater speed during this time means that it is 780 Earth days (2.1 Earth years) between the times of closest approach to Earth.

WHEN MARS is closest to the Earth, a telescope with a magnification of 100 times will show it to be as large as the Moon seen with the unaided eye.

AT ITS CLOSEST approach, Mars is 56 million km from Earth. When it is furthest away, which means on the opposite side of the Sun from us, it is 400 million km away.

THE TWO Mars Exploration Rovers, called Spirit and Opportunity, gradually lose power due to the buildup of Martian dust on the solar panels that provide them with electricity.

Below: After 330 Martian days of exploring the surface of the planet, NASA's Mars Exploration Rover Opportunity encountered its own heat shield debris. The dark disturbed soil marks are where the shield first hit the surface after being jettisoned during entry.

of lava on one half of the Martian surface. This created stresses that literally pulled the sides of the canyon apart, much like opening the top of a muffin by pulling it apart at opposite edges. Other processes, such as landslides and the collection of windblown material, then altered the region further. In the canyon walls, exposed water ice may have turned into water vapour, resulting in the collapse of the surface.

THE ABUNDANCE OF WATER ICE on Mars appears to relieve one of the many, many difficulties of sending humans there – at least they will not need to bring all of the water they need with them if they attempt to land on the planet.

While a manned mission to Mars might seem to be the next logical step for human space exploration, we should not underestimate the difficulty of undertaking such a journey. Even at its closest approach to Earth, Mars is still 150 times farther away than our Moon.

The orbit that would allow the greatest amount of baggage to be brought along for a given amount of fuel, also takes the longest time – 259 days one way! Space is a very unforgiving place and the likelihood of fatalities is practically certain for such a long mission. Thankfully, a great deal can be learned by robotic exploration as human life should not be wasted on missions that can be completed just as easily by machines.

NASA/JPL/CORNELL

IF THE MASS of all the planets were combined, Jupiter would make up 77% of that mass.

AFTER THE SUN, Moon, and Venus, Jupiter is the next brightest object in the sky.

JUPITER'S MOON Ganymede is larger than Mercury and Pluto!

JUPITER IS so large that it could contain over 1400 Earths.

AT LEAST 60 small moons are known to orbit Jupiter.

JUPITER orbits the Sun at 13km/sec (47,000 km/h).

JUPITER contains 318 times the mass of the Earth.

FIFTH FROM THE SUN

JUPITER
the Biggest Planet

Jupiter is HUGE, and very unlike any of the planets closer to the Sun. Mercury, Venus, Earth and Mars all have solid surfaces, but Jupiter doesn't!

THE ENTIRE PLANET is made up of clouds of gas that get thicker and thicker towards the centre of the planet. The gas is mainly hydrogen and helium – the two lightest and most common atoms in the Universe. Jupiter has the same makeup as stars, like our Sun.

The gravity associated with Jupiter's large mass is an important influence on the outer solar system. Comets that return frequently, such as Halley's Comet, are believed to have had their orbits altered by a close encounter with Jupiter sometime in the past. In 1994, a comet that had passed close to Jupiter on an earlier orbit was televised plunging into the planet! It is through many such collisions that Jupiter grew to its current size. The magnetic field that surrounds Jupiter at its cloudtops is ten times stronger than the Earth's magnetic field.

When looking at Jupiter through a small telescope, it is possible to see different coloured cloud

The image of the surface of Europa by the planetary probe Galileo reveals the intricate patterning of the ice on the moon's surface. The rarity of craters indicates the surface is relatively new.

bands and a large storm-like feature called the "Great Red Spot".

When the Great Red Spot is visible, it is easy to see how quickly the planet rotates. Planetary probes have discovered other features of Jupiter, such as lightning and the equivalent of "Aurora Australis" (the southern lights).

JUPITER HAS OVER SIXTY MOONS, but the four largest and brightest – Io, Europa, Ganymede, and Callisto – are called the "Galilean Moons". These four moons played an important role in the history of science. They can easily be seen with binoculars and in 1610 the Italian scientist Galileo spotted them through his small telescope and recognised that they orbited Jupiter. At the time, people incorrectly believed that all of the planetary bodies orbited the Earth.

JPL/NASA

A full hemisphere of Jupiter's moon Io is seen in this image from the Galileo Orbiter. The dark spots mark the centres of recent volcanic activity. Of the solid bodies in our solar system, Io's surface changes the most quickly.

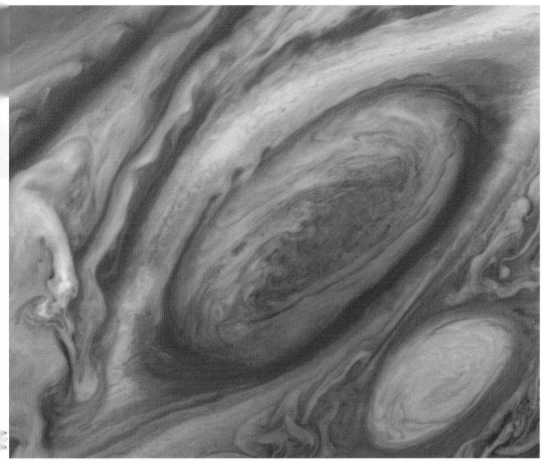

Jupiter's most prominent feature, the Great Red Spot, is seen in this image taken by NASA's Voyager 2 probe. The gaseous and continuously changing nature of this region is apparent. The clouds within the Great Red Spot rotate once within its boundaries in about six days.

IT TAKES JUPITER 11.86 Earth years to orbit the Sun.

ONE DAY on Jupiter takes just 9.9 Earth hours.

JUPITER IS about 780 million km away from the Sun, 5.2 times the Earth–Sun distance.

COMET SHOEMAKER-LEVY 9 collided with Jupiter In 1994.

This is a true-colour image of Jupiter taken by the Cassini spacecraft as it passed Jupiter on its way to Saturn. The most obvious feature is the "Great Red Spot" which has been present for 400 years – and possibly much longer. The different colours are related to the heights of the cloudtops.

Galileo's discovery was the first of many that helped establish our current understanding of the solar system.

Three of the Galilean Moons are larger than our Moon and the fourth, Europa, is only slightly smaller. Europa is thought to have liquid water not far below its icy surface. The moon closest to Jupiter, Io, has numerous volcanoes on its surface. The others are made up of a mix of ice and rock. They would all be bright enough to see with the unaided eye if they weren't so close to the much brighter Jupiter.

The brightness of Jupiter in the night sky is remarkable when you think of how very distant it is. Jupiter outshines the much closer Mars most of the time for two reasons – Jupiter's cloudtops reflect a large fraction of the light they receive, and it is so much larger. The gas and cloudtops at different latitudes on Jupiter don't complete one day in the same amount of time. So features in the cloudtops move with respect to each other as time goes by. The winds at latitudes close to Jupiter's equator are typically over 500 km/h. Closer to the poles they are only about 140 km/h.

This 26 km wide crater on the surface of Europa is believed to be one of the youngest on the entire moon. Such recent impacts expose material from below the surface – in this case, water ice.

UNLIKE THE EARTH, Jupiter is inclined very little to its orbit – only 3.1 degrees.

***

THE AVERAGE cloudtop temperature is a bitterly cold -148°C.

***

THE AVERAGE density of Jupiter is just 24% of the Earth and only 1.3 times greater than the density of water.

***

SUNLIGHT ON the Earth is 27 times more intense than it is at Jupiter's great distance from the Sun.

***

BEFORE THE INVENTION of the sea-going chronometer – a very accurate clock – the times at which Jupiter's Galilean moons entered or emerged from Jupiter's shadow were used to establish geographic longitude on Earth.

***

THE INNERMOST known moon of Jupiter is called Metis. It orbits within Jupiter's ring system and is only 40 km across. Metis was discovered in 1979 in images returned from NASA's Voyager 1 mission.

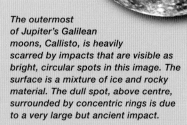

The outermost of Jupiter's Galilean moons, Callisto, is heavily scarred by impacts that are visible as bright, circular spots in this image. The surface is a mixture of ice and rocky material. The dull spot, above centre, surrounded by concentric rings is due to a very large but ancient impact.

The surface of Jupiter's moon Io has numerous active volcanoes. Fresh lava flows on Io are visible in this image from the Galileo spacecraft. This view is about 250 km across.

JUPITER HAS THE STRONGEST

magnetic field of any planet in the solar system – 14 times stronger than the Earth's. Like on our planet, Jupiter's magnetic poles are not in the same locations as the "geographic" poles. So the whole magnetic field pattern is tilted at 9.6 degrees and is swept around once per 9.9 hour rotation. The field traps high-energy electrons and protons, which can impact the atmosphere near the pole of the planet to produce aurora, but also affect the innermost moons, including Io, Europa, and Ganymede.

The surface of Io is actually eroded by these tiny particles. At the same time, the atoms from the surface of Io that are eroded are a source of the electrons and protons surrounding Jupiter! The regions containing these particles are so large that if it were possible to make them visible, you could see the shape with the unaided eye. Sometimes space isn't quite as empty as it first seems.

The particles in Jupiter's magnetic field are travelling at speeds of about 1800 km/h. They are abundant enough that they pose a hazard to electronic equipment in space probes. Furthermore, any form of life which passed through this region would receive a lethal dose of radiation, which obviously limits human exploration of the innermost moons. Since Europa is believed to have a deep layer of liquid water below its icy exterior, these trapped high-energy particles may prevent us from exploring one of the few possible locations for the development of life in our solar system.

The temperature at the centre of the planet is believed to be about 25,000°C and the pressure is 100 million times the atmospheric pressure at the surface of the Earth! These extremes of temperature and pressure near the core of the planet produce conditions where hydrogen gas actually acts like a metal. This is the region where Jupiter's magnetic field is created.

The third-closest of Jupiter's large moons, Ganymede, has a surface largely composed of water ice mixed with material rich in the common element silicon. The surface is evolving over time, but over a much longer period than Io's. Bright spots are fresh water ice in relatively recent impact craters. It is the largest moon in the solar system and is believed to have a large rocky core.

JUPITER HAS A RING SYSTEM but the

particles are so fine that it is only possible to see them if you are looking towards the Sun from behind Jupiter. They are thought to be partly the result of material from the volcanoes on the innermost large moon, Io. Unlike those of Saturn, Jupiter's rings are not made of ice.

Many spacecraft en route to more distant planets have taken advantage of Jupiter's strong gravity to alter their trajectories and add speed in the direction of their destination. This way of saving rocket fuel is called a "gravitational slingshot".

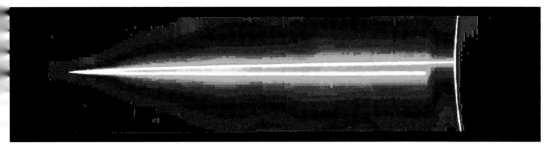

Jupiter's rings are best seen when they are backlit by the Sun, since they are composed of very small particles. This false-colour image mosaic was built up from images taken by the Galileo planetary probe. Jupiter is seen in silhouette on the right. The rings appear to end abruptly due to the shadow of Jupiter.

The probes often take advantage of this encounter to test out their imaging system and other scientific instruments on Jupiter.

Two probes from Earth have entered Jupiter's atmosphere. The first was an atmospheric probe released by Galileo designed to study the material that makes up Jupiter's cloudtops. The second was the Galileo space probe itself. The power source for the craft was a module containing highly radioactive uranium. Rather than risk the possible future collision of the probe with a moon in space – possibly even a moon with useful water on it below the surface – it was decided to "ditch" the spacecraft by deliberately colliding with Jupiter.

IS JUPITER ALMOST A STAR?

Jupiter and its Great Red Spot as it appears through a small telescope. While not as detailed as the images from planetary probes, it is still possible to follow the changes of the cloudtops from night to night and to witness the changing positions of Jupiter's Galilean moons as they orbit.

ROBERT BOTTS

EARLIER IN THIS BOOK we learned that the properties of a star are determined by how much gas is collected together by gravity, known as the mass. In this chapter, we learned that Jupiter's most common elements are the same as those of stars and in almost the same proportions.

So, how close is Jupiter to producing its own energy through nuclear fusion at its centre? In other words, how close is it to being a star?

The answer is "not very close". Jupiter contains about $1/60th$ of the mass it would need to produce significant amounts of fusion energy.

However, Jupiter does give off about 60% more energy that it receives from the Sun. How could that be if it isn't producing its own? The answer is interesting. Objects falling onto Jupiter deposit all of the energy of their motion into its gas. Since objects are moving very rapidly when they hit Jupiter, and since the planet has been built up entirely by these collisions, a lot of heat energy is present in the bulk of the planet. The only way to lose this heat to space is through Jupiter's cloudtops. Unfortunately, a round object is the least efficient shape for losing energy this way, and the large amount of mass stores heat until it can be leaked away into space.

A second factor is that as more material falls on Jupiter the gas below is compressed even more, generating even more heat.

Jupiter won't add significantly more mass in the aeons ahead, except for the occasional object like Comet Shoemaker-Levy 9.

When enough time has passed, Jupiter will eventually emit only as much energy as it receives. In the meantime, Jupiter doesn't qualify as a star but it isn't your average planet either!

ON ITS ORBIT, Jupiter comes closest to the Earth every 399 days, but even then it is still about 590 million km away. Its greatest distance from the Earth is just below 1 billion km.

THE MAGNIFICATION required for a small telescope to show Jupiter as the same apparent size as the Moon (to the unaided eye) is only 40 times.

JUPITER TRAVELS 4.8 billion km during one orbit of the Sun.

TO HAVE ENOUGH SPEED to escape from the cloudtops of Jupiter, you would need to be moving at over 200,000 km per hour.

NASA AND THE HUBBLE HERITAGE TEAM (STSCI/AURA)

The Hubble Space Telescope captured this ultraviolet image of aurora on Jupiter. The Galilean moons affect Jupiter's magnetic field enough that bright spots in this image can be traced back to the moon responsible!

THE DIFFERENCES in positions of the Galilean moons orbiting Jupiter are easy to see from night to night using binoculars or a small telescope. When all four aren't seen, it is because one or more are in shadow, crossing the face of Jupiter, or behind it.

OLE ROMER, at the Paris Observatory in the 1670s, noticed that the predicted times of eclipse events for the Galilean moons were dependent on the Earth's orbit relative to Jupiter. He correctly found that a finite speed of light would explain discrepancies.

Above: This image of Saturn and its intricate rings was produced from 126 images from the Cassini spacecraft. Note how Saturn is visible through the rings and how the rings are not a solid disk. This partially-illuminated view of Saturn is never seen from Earth.

NASA/JPL/SPACE SCIENCE INSTITUTE

SIXTH FROM THE SUN

SATURN
with Beautiful Rings

Aside from our home the Earth, Saturn is without a doubt the most beautiful planet because of the extensive rings that surround it.

A SMALL TELESCOPE will reveal Saturn and its remarkable rings, which appear to be solid but are actually made of separate chunks of ice, most smaller than 1 m across. The rings are 300,000 km across but only about 1 km deep.

NASA/JPL/SPACE SCIENCE INSTITUTE

WHY DOES
SATURN HAVE RINGS?

When small pieces of ice or rock are orbiting a planet, they usually clump together and form a moon. The bigger the moon, the more easily its gravitational pull attracts material. Strangely, gravity can rip apart moons that are too close to a planet by exerting a stronger force on the side near the planet than on the other side. Ring systems, such as Saturn's, are the result. Saturn is a "gas giant" planet much like Jupiter, although slightly smaller. Over 99% of the planet is made up of hydrogen and helium.

Other gases found on Saturn are methane and ammonia. Like Jupiter, Saturn has no solid surface, only cloudtops. Bands of slightly different colour are visible in a telescope, but there is no counterpart to Jupiter's Great Red Spot. If it weren't for the fascinating rings, Saturn would seem a less interesting version of Jupiter.

The brightness of Saturn in the night sky depends mainly on how much its rings are tilted in our direction: the more the tilt, the brighter it appears. Every 15 years, when Saturn's rings are almost edge-on, the planet becomes harder to see, Mostly, Saturn is as visible as the brightest stars.

Below: This image of Saturn's rings covers about 62,000 km. The ring system is very complex. Gaps and patterns in the rings are due to the gravitational pull of Saturn's many moons. Each ring is made up of countless individual chunks of ice.

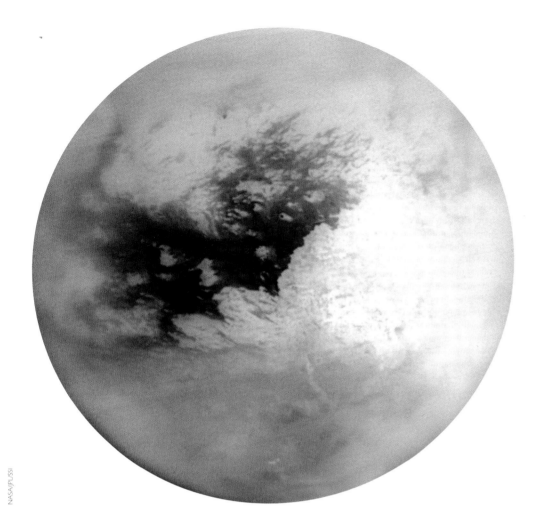

NASA/JPL/SSI

This full-disk image of Titan was pieced together from 16 images obtained by the Cassini probe, which is in orbit around Saturn. At visible wavelengths Titan appears featureless due to the layer of haze. However, Cassini recorded the images in the infrared where Titan's atmosphere is more transparent. The nature and origin of the surface features is still under debate. The near-complete absence of impact craters argues for an active, young surface. Erosion due to drainage of liquid was clearly visible in images from the Huygens probe as it descended onto Titan in 2005.

STRANGE MOONS

There are over 30 moons orbiting Saturn. The largest of these is called Titan. Saturn's other moons are mostly made of ice and many of the smaller ones are probably comets that have been attracted by the planet's gravity.

TITAN is the most distant object on which a planetary probe has landed. In January 2005, the European probe Huygens touched down on Titan's surface, which was 1.2 billion km away from Earth at the time. Despite the surface temperature of -180°C, the lander relayed data for two hours before succumbing to the chill.

Huygens took many images as it descended towards the surface of Titan and also recorded the "wind" noise in the atmosphere. The images reveal what seem to be drainage channels that flow towards a lower-lying dark "ocean", presumed to contain some kind of tarry substance. They also showed a fog bank, apparently made of condensed methane.

The colour of the sunlight that has penetrated the atmosphere at the surface of Titan is a distinctive orange. Even before sunlight enters Titan's atmosphere it is 1/100th the brightness of the Sun at Earth, due to Saturn's (and Titan's) much greater distance. The illumination at the surface of Titan is never greater than the brightness of twilight on Earth.

NASA/JPL/SPACE SCIENCE INSTITUTE

The upper atmosphere of Saturn's largest moon, Titan, is seen in this image taken by the Cassini spacecraft. The thick haze containing methane prevents the surface of Titan from being seen directly in visible light. Only about 10% of the light hitting the top of the atmosphere reaches Titan's surface.

LIKE MANY SMALL MOONS and objects in the outer solar system, Saturn's moon Hyperion probably contains many interior gaps, since its gravity is not sufficient to crush material together tightly.

UNDERNEATH TITAN'S haze layer, clouds have been detected by the Cassini probe and ground-based telescopes. Unlike Earth's clouds, they do not cover the entire planet but form only at certain latitudes. The location of these clouds seems to depend on the Saturnian season.

Saturn's moon Mimas, taken by the Cassini spacecraft, is just under 400 km in diameter, but has a crater that is 140 km across. Had the object that made this crater been larger, it would have broken Mimas into small pieces.

The moon Hyperion is irregularly shaped at 328 x 260 x 214 km. It tumbles instead of spinning.

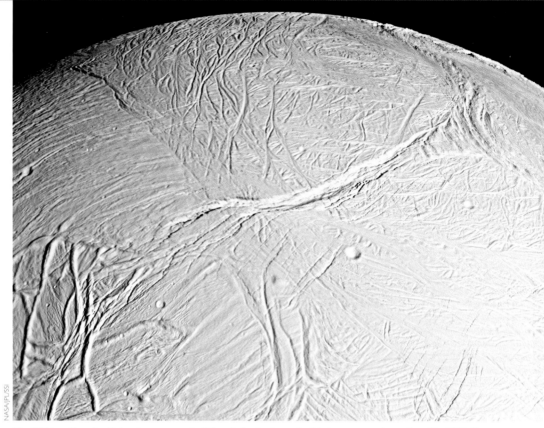

The 505-km diameter Saturnian moon called Enceladus is shown in this image mosaic captured in 2005. Numerous geological features on the ice crust of this moon suggest the build up and release of stresses. Recent imagery has revealed sprays of fine water ice particles emanating from the surface and reaching altitudes as high as 250 km.

RHEA, the second largest of Saturn's moons, is also visually interesting. Like many moons in the solar system (including our own Moon), its rotation has been locked by the effects of Saturn's gravity so that it only rotates once per orbit. Saturn only ever sees one side of Rhea. The side facing into the orbit is full of craters, but the other side also has bright, streaky features, which may be ice cliffs, that run for half the moon's circumference. Like many of Saturn's moons, Rhea is made mostly of water ice.

IAPETUS, Saturn's third-largest moon, with a diameter of over 1400 km, is in some ways the most peculiar moon in the solar system.

It is ten times darker on one side than it is on the other! The source of this surface difference is still uncertain.

Because it rotates once for every orbit of Saturn, the same side, the darker side, is always facing forward along the orbit. This has led some astronomers to believe that it is dark because it has picked up material along its orbit, but there seems to be more dark material in crater floors than elsewhere, which suggests the source may be inside the moon! The fact that the craters on the darker side are not whitish reflective ice means that either the dark material is quite thick or that it is constantly replenished so that even fresh craters are quickly darkened.

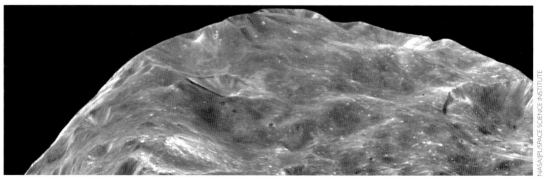

Saturn's 120 km diameter moon Phoebe is seen in the image mosaic taken by the Cassini probe from a distance of only 8,000 km. The brightest spots are probably ice. Some of the crater walls are as high as 4 km.

URANUS, NEPTUNE,
& Beyond

Uranus, Neptune and the "dwarf planet" Pluto are so far from the Sun that they intercept few of its warming rays. None of them are like our Earth – they consist mainly of hydrogen and helium.

PLANET URANUS

Uranus was the first planet discovered with a telescope. It was not known to ancient people because it is barely bright enough to be seen with the naked eye, even under the very best conditions. The English astronomers William and Caroline Herschel found Uranus while "sweeping" the sky for interesting objects in 1781. It had a clearly visible disk of a blue colour, and moved from night to night. Initially, they thought they had found a new comet. Six years later, in 1787, William Herschel also discovered two of the planet's brightest moons, which have been named Titania and Oberon.

Uranus is the third largest planet. It was also the second planet discovered to have rings. It is chiefly made up of hydrogen (83%), helium (15%) and molecules composed of hydrogen and carbon or nitrogen. An usual aspect of Uranus is that it rotates on its side. This has the remarkable effect that portions of the planet are either bathed in sunlight or completely dark for decades during its 84-year orbit of the Sun.

PLANET NEPTUNE

Neptune is slightly more massive but slightly smaller in size than Uranus.

In 1613, Galileo accidentally recorded an observation of Neptune, thinking it was a star, when it passed near Jupiter. He even noted that it seemed to have changed its position slightly relative to another star.

It has the remarkable distinction that it was predicted to exist before it was actually seen, based on changes in the orbit of Uranus which could not be accounted for

NASA/JPL

Above: The pale blue and nearly featureless disk of Uranus is seen in this true-colour image from NASA's Voyager 2 probe in 1986. Fortunately, the moons orbiting this planet turned out to be much more visually interesting.

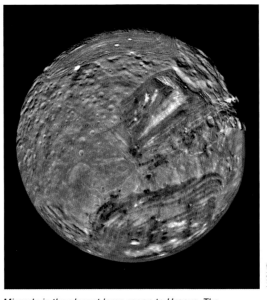

NASA/JPL

Miranda is the closest large moon to Uranus. The moon appears to be half ice and half rock. The abrupt changes in its surface features have led astronomers to suggest that it was broken apart by an impact and the pieces reassembled themselves but not in their original orientations. Its diameter is 470 km.

THE FACTS!

MANY OF the smallest moons orbiting the planets Jupiter, Saturn, Uranus, and Neptune are believed to be captured comets.

NEPTUNE APPEARS to have a blue cast due to red light being absorbed by the atmospheric gas methane.

URANUS AND Neptune rotate once in 17.25 and 16 hours, respectively.

THE COMPOSITION of the Neptunian atmosphere is much like the other gas giant planets and the Sun. About 80% is a molecule consisting of two hydrogen atoms, 19% is helium, and 1.5% is methane.

THE INTERIOR of Neptune contains heavier substances such as molten rock and metals. There is also liquid water and ammonia at high pressures and temperatures.

SMALL MOONS continue to be discovered orbiting the outer planets. At last count, Uranus has a total of 27 moons and Neptune has 13.

THE AMOUNT OF LIGHT reflected by Neptune is 29%, which is also very similar to the Earth's average of 31%.

THE WINDS in Neptune's atmosphere are ferocious at up to 2000 km/h.

THE SURFACE GRAVITY on Neptune is very similar to the Earth's – you would weigh only 14% more on Neptune.

THE LARGE MOON TRITON orbits Neptune in the opposite sense to the planet's rotation. This continuously reduces the size of Triton's orbit, which will eventually cause Triton to be ripped apart by tidal forces.

NEPTUNE HAS a magnetic field, but it is highly tilted and centred halfway to the surface from the centre of the planet.

CHARON is roughly half the diameter of Pluto, but it was unknown until 1978 when an astronomer noticed that some images of Pluto appeared elongated – even though all the surrounding star images were round.

by the known planets. In 1845 and 1846, the English astronomer Adams and the French astronomer Leverrier independently analysed the deviations of Uranus and predicted the location where the new planet could be found. The German astronomers Galle and d'Arrest received the predictions by mail and searched for and found the planet the very next evening! Neptune is too faint to be located with the unaided eye, but can be seen with almost any pair of binoculars if you know where to look.

Neptune orbits the Sun with a speed of 20,000 km/h but takes 165 Earth years to orbit the Sun once. It is 30 times further away from the Sun than the Earth and therefore receives only $1/900$th the light.

There is a system of rings around Neptune. Unlike those around Saturn, they are relatively sparse and cannot be seen directly from Earth. The ring material is not very reflective. It is much darker than Saturnian ring ice.

If the interpretation of recent data is correct, the rings may be quite short-lived features of the Neptunian system. Only one planetary probe has visited Neptune – NASA's Voyager 2. It flew past the planet, its rings, and moons in 1989. Only two moons were known prior to the encounter. Six new moons were located by Voyager 2.

NASA/JPL

Above: Neptune imaged by the space probe Voyager 2. The feature named the "Great Dark Spot" is similar to its counterpart the "Great Red Spot" on Jupiter. The bright features come and go with time, indicating a dynamic atmosphere.

DWARF PLANET PLUTO

Pluto was found in 1930 by the American astronomer Clyde Tombaugh, who had been given the task of trying to find a ninth planet using the techniques that had worked in the search for Neptune.

In hindsight we realise that the discovery of Pluto was a fluke. Its mass, at just 0.2% the mass of the Earth, is now accurately known to be far too small to have affected the orbit of Neptune. The density of both Pluto and Charon, one of the dwarf planet's moons, is about twice that of water, indicating that it must be composed of both ice and rocky material. The rotation pole is highly inclined relative its orbit, resulting in long periods when a single pole faces

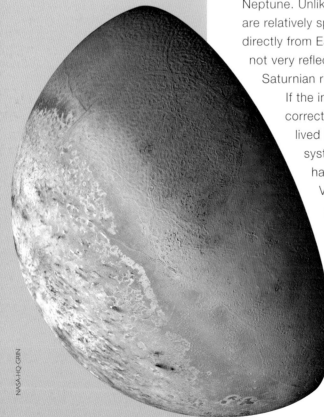

NASA-HQ-GRIN

The surface of Triton, Neptune's largest moon, captured by Voyager 2. Triton is 1350 km in diameter, smaller than the Earth's Moon. However, its surface is much more interesting. It has a thin atmosphere of nitrogen that condenses as frost. The dark streaks are from geysers which expel gas from the interior. The surface is extremely cold at -235°C.

This is one of the best images of Pluto and the "moon" which orbits it, called Charon. Both objects are believed to be comet-like in composition. Details on either surface cannot be seen. Pluto and Charon were 4.4 billion km away when this image was taken. Pluto and Charon are 2320 and 1270 km in diameter, respectively.

THE ORBIT of Pluto is elongated enough that it is sometimes closer to both the Sun and the Earth than Neptune.

THE SURFACE TEMPERATURE on Pluto is between -235 and -210°C.

CHARON ORBITS the dwarf planet Pluto in 6.4 Earth days, the same length of time it takes both Pluto and Charon to rotate once. The same sides of Pluto and Charon always face each other.

the Sun or is in shadow. Present images of Pluto reveal little detail, but it does appear that the contrast of different surface areas is greater than any of the eight major planets.

The orbit of Pluto is more elliptical than any of the eight large planets in the solar system. It is tipped 17 degrees away from the orbit of the Earth – greater than any of the eight major planets and more than twice as much as Mercury's orbital inclination (7 degrees). Pluto's orbital period around the Sun is exactly 50% greater than Neptune's. This relationship is called a "resonance". Even though Pluto is sometimes closer to the Sun than Neptune, the resonance ensures that the two bodies are always well separated from each other at those times.

Pluto is smaller than several moons of the major planets! It takes 248 Earth years to orbit the Sun. At its closest, it is

4.4 billion km from the Sun, and at it is farthest it is 7.4 billion km away.

A space probe called "New Horizons" was launched in January 2006 to fly past Pluto and its moon Charon. It will provide the first high-resolution images of these mysterious bodies and answer key questions about the formation and evolution of similar objects in the outer solar system. New Horizons will use a gravitational assist by Jupiter in early 2007 to gain speed and is scheduled to arrive at Pluto in July 2015.

After a journey of 5 billion km, the probe will pass within 10,000 km of Pluto and 27,000 km of Charon. At its closest approach, it will be travelling at 50,000 km/h – hundreds of times faster than a jet! Most of the images will be taken during a single Earth day. Pluto is so distant that one-way travel time for the radio waves from New Horizons will take 4 hours and 25 minutes!

The Hubble Space Telescope recently discovered two new objects orbiting Pluto, bringing the number of the dwarf planet's "moons" to three. These images were taken three days apart in May 2005. The change in pattern is due to the moons moving along their orbits. Pluto is shown as yellow and Charon as blue. The new objects have been named Nix and Hydra.

WHAT IS A DWARF PLANET?

ASTRONOMERS HAVE KNOWN for many years that Pluto didn't belong in the same class as other planets. Recently, additional Pluto-like objects have been found at even greater distances from the Sun. One of them, Eris (formerly known as 2003 UB313), appears to be at least as large as Pluto.

The International Astronomical Union, at its triennial meeting in Prague in August 2006, adopted a new class of object called a "dwarf planet" and identified Pluto as its first example. The most important characteristic that sets dwarf planets apart from normal planets is their low mass, which is not enough to "clear" the orbital region through the effect of the dwarf planet's gravity.

EVENTS IN THE NIGHT SKY

SHOOTING Stars

The appearance of a "shooting star" or meteor in the night sky is exciting and startling – most celestial events happen much more slowly.

SHOOTING STARS appear to fall out of the sky, but the reality is very different. What you see is taking place in the Earth's atmosphere, not in outer space, and the object causing it is smaller than a pebble!

Meteors come from two sources, both involving objects orbiting the Sun that happen to cross into the Earth's orbit. The first kind is a rock or pebble that was once part of a larger object. A collision with another object, or the gravitational pull of a large planet such as Jupiter, throws these small rocks into a different orbit. These types of meteor happen all the time but are best seen on clear, moonless nights. You may count around six shooting stars an hour if you have the patience to sit and watch. The incoming rock ploughs into the Earth's atmosphere at around 75,000 km/h. Due to air friction, it heats up and slows down in a fraction of a second. The hot air around it glows and expands as the object burns up. You don't see the hot pebble – you see the

1cm

This meteorite from the Antarctic ice sheet was once part of the crust of Mars. It was blasted off by an impact 16 million years ago and fell to Earth about 13,000 years ago.

glowing path of heated air! Pebble-sized objects burn up completely, but rocks the size of cricket balls are large enough to survive the trip and sometimes fall to Earth – they are then called meteorites. Once every ten thousand years or so, a meteorite large enough to produce a crater hits the Earth. Dry regions of Australia with low rates of erosion are the best places to locate meteorites and view old craters.

The other kind of meteor comes from a comet. Comets are objects made mostly of ice and rock-like particles. Comets that travel close to the Sun lose some of their ices with each passage, leaving only rock-like particles which gradually disperse along the comet's orbit. If the orbit of an old or dead comet crosses the orbit of the Earth, we pass through the particle debris on the same dates each year, giving rise to what is known as a "meteor shower". There is no known case of a meteor from a comet surviving its trip through the atmosphere.

NASA/JPL/CORNELL

While exploring the surface of Mars, NASA's Mars Exploration Rover Opportunity located this meteorite – the first ever seen on another planet.

MOONS AND STARS

ECLIPSES
of the Sun & Moon

Rare events like eclipses stand out in the slowly changing daytime and night-time skies.

A SINGLE LOCATION on Earth can go for months or years between witnessing solar or lunar eclipses – longer if poor weather is factored in. Since the Sun and Moon are the most obvious objects in the sky, it is not surprising that eclipses have a long history of terrifying people in pre-scientific times. There is a known instance of ancient Chinese astronomers being executed for failing to predict a solar eclipse! Babylonian astronomers noticed that eclipses could be predicted to occur 18 years and $11^{1}/_{3}$ days after previous eclipses. The one-third of a day remainder means that the path of totality for a solar eclipse won't occur in the same place, but a partial eclipse would be visible, weather permitting!

One of the first clues that the Earth was round was the circular profile of the shadow of the Earth during a lunar eclipse. The ancient Greeks understood how lunar eclipses happened and correctly concluded that the Earth must be spherical.

HOW ECLIPSES OCCUR

Light from the Sun heads away from it in all directions. Any solid body in the solar system will cast a shadow away from the Sun. If another object enters that shadow, it is said to be "in eclipse". For the Earth–Moon system, there are two alignments that can produce eclipses. When the shadow of the Moon falls on the surface of the Earth, it is called a solar eclipse. When the Moon enters the Earth's shadow, a lunar eclipse

occurs. Solar eclipses happen at New Moon and lunar eclipses happen at Full Moon.

If the Moon's orbit were not tilted with respect to the Earth's orbit around the Sun, there would be one solar eclipse and one lunar eclipse each month. The fact that the Moon's orbit is tilted by just over 5 degrees causes the Moon or Earth to miss the other shadow most of the time.

THE FACTS!

BRIGHT RED loops of material from magnetic storms at the edge of the Sun's disk can be seen during a total solar eclipse.

THE LONGEST total eclipses of the Sun occur in early July, because that is when the Earth is farthest from the Sun and therefore when it appears to be smallest.

SOLAR ECLIPSES occur on all planets with sizable moons orbiting near the planet. Such eclipses on Jupiter are easy to see with a small telescope and occur frequently.

THE WIDTH of the path of totality during a total solar eclipse is rarely greater than 250 km on the Earth's surface.

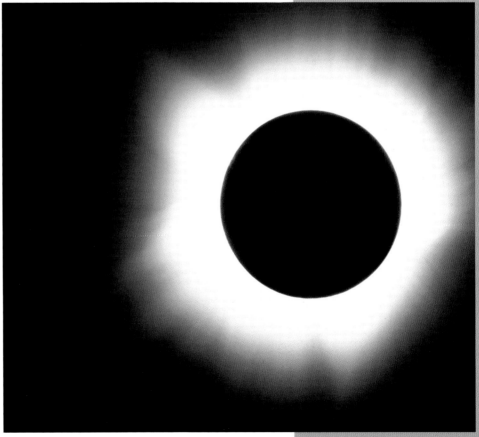

This image of the totally-eclipsed Sun reveals its much fainter outer atmosphere called the corona. The gas in the corona has a temperature of millions of degrees Celsius and is only visible to the naked eye in the path of totality during a total solar eclipse.

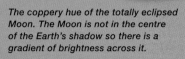

The coppery hue of the totally eclipsed Moon. The Moon is not in the centre of the Earth's shadow so there is a gradient of brightness across it.

This sequence of frames shows the dark disk of the Martian moon Phobos eclipsing the Sun, as seen by NASA's Martian Exploration Rover Opportunity from the surface of Mars. This event would correspond to an annular solar eclipse on Earth, but our Moon is much more similar to the Sun in size.

WHEN DO ECLIPSES OCCUR?

There are two periods during each year when eclipses can occur, and these are separated from each other by six months. The Moon orbits the Earth 13.0357 times during a calendar year, so the months when eclipses can occur change from year to year.

SOLAR ECLIPSES

A partial solar eclipse occurs when the observer on the Earth is at a position where the Moon only partially covers the disk of the Sun. For any given place, partial eclipses will be visible more frequently than any other kind. People's eyes adjust easily to slow changes in illumination, so the fact that a solar eclipse is occurring may not even be noticed until over three-quarters of the Sun's disk is covered. There is no safe time to look at a partial solar eclipse with the unaided eye.

By complete coincidence, the angle of the Sun and the Moon in the sky is almost the same. The Earth's orbit around the Sun is not perfectly circular; nor is the Moon's orbit around the Earth. Consequently, the apparent size of the Sun and the Moon varies enough that either object can appear slightly larger at any given instant.

This results in two additional types of solar eclipses: annular and total.

When the size of the Moon appears less than that of the Sun, a ring of light from the Sun's disk will still be visible around the Moon. This ring, also called an annulus, will be narrow, but enough light from the Sun's disk will be visible to cause damage to unprotected eyes. The dramatic reduction in the level of sunlight during an annular eclipse frequently tricks birds and other animals into thinking it is dusk.

The most spectacular kind of eclipse is, without doubt, a total solar eclipse. In this case, the Moon's apparent size is larger than the Sun's and, briefly and for a very small region of the Earth, the Sun will be completely covered. (Regions away from this "path of totality" will see a partial solar eclipse.) The duration of the total part of the eclipse is short – often two or three minutes and never longer than seven and a half minutes.

The effect of totality is dramatic – a pitch black hole appears in the sky surrounded by the eerie, outermost layers of the Sun's upper atmosphere. The ambient light during an eclipse comes from a ring of light near the horizon where the distant sky is still illuminated by a partially-eclipsed Sun. Usually there is a significant temperature drop, and often there is a sudden formation of clouds. It all adds up to a very memorable experience indeed!

If you are in the path of totality, during the brief total portion when all parts of the Sun's bright disk are covered, it is safe to glance at the Sun.

Total (and annular) eclipses cover such a small path on the Earth and are so infrequent that it is usually necessary to plan a trip to see one. Many travel companies offer eclipse tours or cruises designed to allow the traveller to experience these spectacular events.

LUNAR ECLIPSES

Lunar eclipses, also called eclipses of the Moon, are everything that solar eclipses are not. They are always safe to view with the unaided eye and everyone in the same hemisphere where the Moon is in the sky can see exactly the same thing! Total lunar eclipses are just as common as partial lunar eclipses.

During a partial lunar eclipse (or the partial phases of a total lunar eclipse) a dark "bite" will appear to have been taken out of the

This sequence of images was taken during the total lunar eclipse of November 8–9, 2003. Clouds cover parts of the uneclipsed and partially-eclipsed Full Moon in the first two images on the left. The remaining three images are taken in totality, the rightmost being when the Moon was deepest in the Earth's shadow.

ROBERT BOTTS

THE SUN is no more dangerous than usual to view during a total solar eclipse.

WHEN THE Moon's shadow just misses the Earth's surface, astronauts in orbit can sometimes observe it.

LONG BEFORE astronauts landed on the Moon, astronomers used the rapid cooling of the Moon's surface during a lunar eclipse to determine that it had a layer of lunar soil on it and was not bare rock.

SEARCHES FOR small objects orbiting the Moon have been made during total lunar eclipses. The Moon does not appear to have any moons of its own!

THE HIGHEST OCEAN tides occur during lunar and solar eclipses because the effects of the Sun and Moon on the ocean add to each other most effectively at these times.

SUPERSONIC AIRCRAFT have been used to stay in the region of totality as the Moon's shadow races across the Earth's surface in order to make scientific measurements.

THE NEXT TOTAL lunar eclipses visible from Australia are on August 28, 2007, and December 21, 2010.

THE NEXT PARTIAL lunar eclipses visible from Australia are on September 7, 2006, August 16, 2008, and December 31, 2009.

Full Moon. This bite will appear on the northern or southern portion of the Moon's disk, depending on the details of the eclipse. A total lunar eclipse happens when the Moon is entirely within the Earth's shadow.

If the Earth had no atmosphere, the Moon would not be visible during this time since no sunlight would reach it or be reflected off its surface. However, the Earth's atmosphere does scatter a small amount of light into the Moon's shadow, so the Moon is visible during totality.

In fact, the Moon usually appears to have a copper hue when completely eclipsed. These are some of the same deep orange and red rays of sunlight that are seen around sunset and sunrise. The darkness of the eclipsed Moon during any particular total lunar eclipse is determined by the amount of dust and smoke in the Earth's atmosphere. After large volcanic eruptions where volcanic dust has been spread widely, total lunar eclipses can be especially dark. To understand what is going on during a lunar eclipse, it is useful to pretend you are viewing events from the surface of the Moon. When the Earth has partially covered the Sun's disk, a partial eclipse is in progress at the location of your lunar vantage point. The Earth's disk is significantly larger than the Sun's in the lunar "sky", so there is a long period when the Sun is totally eclipsed. During this period, you would see a thin, bright, ring of orange light around the Earth where sunlight is being scattered towards your spot on the Moon.

PAUL MORTFIELD

This image of the partially-eclipsed Full Moon reveals the curvature of the edge of the Earth's shadow. When the sunlit portion of the Moon is properly exposed, the dark orange portion within the shadow is not seen.

DID YOU KNOW?

THERE ARE estimated to be around a million million comets orbiting the Sun.

COMETS THAT come too close to Jupiter or Saturn can be thrown into the inner solar system or ejected from it altogether due to the effect of the strong gravitational fields of those very massive planets.

THE MASS of a comet is similar to that of a single mountain on the Earth.

THE DISINTEGRATION of comets is responsible for meteor showers.

THE BRIGHT comet with the shortest orbital period is called Comet Encke and it orbits the Sun once in every 3.3 years.

COMET SHOEMAKER-LEVY 9 collided with Jupiter in 1997. The collision of the largest fragments of the comet – which broke apart during its previous close encounter with Jupiter – released an amount of energy equivalent to 6,000,000,000,000 tonnes of TNT, or roughly 600 times the energy of all of the nuclear weapons in the world.

These Hubble Space Telescope images, starting with the bottom image, show the impact sites of two fragments of Comet Shoemaker-Levy 9 changing over the course of five days.

EVENTS IN THE NIGHT SKY

COMETS
– "Hairy Stars"

A comet that is easily visible to the unaided eye is a relatively rare occurrence.

IN THIS MODERN ERA of well-lit cities and fewer outdoor activities, most people have never seen a comet. Pictures of comets with tails seem to reinforce the mistaken idea that they race across the sky in a very brief period of time. In reality, a comet appears stationary in the sky when you observe it on a given evening or morning, but will have moved noticeably among the constellations from one night to the next.

Throughout history, the appearance of a bright comet was seen as a bad omen – a warning from the heavens of impending disaster. It is only in the last 100 years that we have fully understood comets, where they come from, what they are made of and how they behave.

HOW BIG ARE COMETS?

Apart from the Sun and the Moon, a bright comet is the only celestial object that appears to have a significant size. Even the largest planet, Jupiter, seems to be just a bright point of light to the naked eye. However, comets are not actually very large. The solid centre is typically only 10 km in size and the largest comet observed so far was only 100 km.

WHAT IS A COMET MADE OF?

A comet is sometimes described as a "dirty snowball". Its makeup is largely ices (from water, methane, ammonia and other gases) with a small amount of granular solid particles mixed in. Comets form in the outskirts of the solar system where the temperature is very low.

Planets orbit the Sun in nearly circular orbits, but comet orbits are long and thin. They spend thousands of years (or more!) near the most distant point of their orbit and plunge into the inner solar system near the Sun for only a few months.

Higher temperatures close to the Sun cause the ices to become vapour and the solid particles to be freed from the surface to form a large (about 100,000 km) cocoon around the nucleus. We see this cloud of gas and ice particles and think a comet must be large, but the nucleus is very small.

The bright flash caused by the collision of a probe, released by NASA's Deep Impact mission, with Comet Tempel 1 in July 2005. The after-effects were studied by Earth-based telescopes.

One of the brightest comets in decades, Hale-Bopp, is captured in this image from March 1997. Three exposures have been stacked and shifted to compensate for the motion of this comet across the sky.

ROBERT BOTTS

COMET TAILS

Comets frequently have tails that point away from the Sun. These are formed by the gases and particles released when the comet's surface became warm. The tail can extend for millions of kilometres away from the comet centre, but there is almost nothing to it!

Almost everyone has heard of Halley's comet, which returns every 76 years. It was named in 1782 after Edmund Halley, who first understood that comets orbited the Sun and could reappear at regular intervals. In 1912, the Earth passed through the tail of Halley's comet with no ill effects – but it did create a spectacular meteor shower!

In recent years, space probes have flown very close to the nucleus of several comets. The images that have been sent back to Earth reveal that these bodies are indeed as small as we expected and that they also have irregular shapes. When sunlight interacts with some parts of the comet's surface, high-speed eruptions of gas occur. These can be powerful enough to alter the trajectory of the comet – a little like having rocket thrusters!

After enough encounters with the Sun, the comet disintegrates. It loses all of its ices and can release no more material. The stripped comet continues orbiting the Sun indefinitely, but it will never again be visible to a skywatcher.

SPEEDY COMETS

THE MINIMUM RELATIVE SPEED of the Earth and a comet passing close is 44,000 km/h. The maximum relative speed at which a comet and the Earth could collide is 260,000 km/h.

At the Earth's distance from the Sun, objects in a circular orbit move at 108,000 km/h and objects in a cometary orbit move at 152,000 km/h. If the directions of these motions are opposite each other, then the maximum relative speed is achieved.

Most often, however, the directions of motion won't be so perfectly aligned and the relative speeds will be somewhat less.

THE FACTS!

SHORT-PERIOD comets like Halley's comet have orbits that are little inclined to the orbits of the major planets. Long-period comets can approach the Sun from any direction.

SATELLITE INSTRUMENTS studying the Sun have revealed that many small comets exist but only become easily visible when they are closest to the Sun. A number of comets have been seen to disappear/ evaporate as they are heading on a collision course with the solar surface.

THE VAST majority of cometary objects orbit at huge distances from the Sun and never enter the inner solar system. Their small size when inactive makes them difficult or impossible to detect at distances beyond Neptune.

THE SOLID portion of the comet – its nucleus – is not strongly held together. It is fairly common for a nucleus to break up into multiple pieces when it is near the Sun (or gets too close to Jupiter!) Since a fragmentation produces a host of smaller objects with more combined surface area, comet breakups usually lead to dramatic brightening and the possibility of multiple tails.

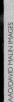

AAO/DAVID MALIN IMAGES

Comet Hyakutake is seen in this colour-composite image taken in March 1996. Stars appear as red, green, and blue dots because the telescope was shifted between exposures to follow the comet.

THE PART of the atmosphere where aurorae occur is between 100 and 500 km altitude.

THE COLOUR of an aurora depends on how deep into the atmosphere the particles penetrate.

ALTHOUGH AURORAE can occur at any time of night, the conditions are most favourable around the hours of local midnight. Aurorae can last only a few minutes or all night.

THE EARTH'S MAGNETIC FIELD is changing strength and position and this affects where and how often aurorae are seen. The strength of the field in 2000 was only 83% of what it was in the year 1600.

DURING AURORAL displays, huge electrical currents flow in the higher layers of the atmosphere.

THE RAPIDLY CHANGING magnetic field during a major aurora can cause power transmission lines to fail and "black-out" urban regions. These black-outs allow city-dwellers to see aurorae under country sky conditions!

RADIO COMMUNICATIONS can be disrupted by the effects of geomagnetic storms associated with an aurora.

MAGNETIC disturbances during aurorae can lead to corrosion in high-latitude oil pipelines in Alaska, Canada, and Siberia!

Right: The brightest auroral displays can be seen even in the city in conditions of strong light pollution. Such aurorae are frequently very active and cover a large fraction of the sky.

THE SOUTHERN LIGHTS
AURORA
Australis

Displays of "southern lights" are a rare occurrence, but they are so unusual and so beautiful that they are worth looking for. Very few people in modern cities have seen them.

WHAT DO THEY LOOK LIKE?

At its best, the Aurora Australis resembles transparent green or reddish curtains of light shimmering and swaying in the sky. They sometimes appear to flicker like a flame, all over the sky at once. Two more common forms are a long shallow arc above the southern horizon or a vague patch of coloured light, which is most likely where the name "aurora", meaning "dawn", came from.

ROBERT BOTTS

WHEN CAN THEY BE SEEN?

Unlike many things in the sky, there is no season when aurorae occur more frequently although the earlier evenings of the winter do give people more opportunity to notice them. The appearance of aurorae is determined by how active the Sun's surface is and this is related to how many large sunspots appear there.

A news report of a large solar flare is a good indication that aurorae are possible roughly 48 hours later.

There are numerous websites on the internet which provide nearly real-time predictions of auroral activity. The best views of these curtains of light occur far from city lights, when the Moon is not in the sky, but the very brightest displays can be seen even from the centres of large cities.

Aurorae are most frequently seen during the peak of the 11 year sunspot cycle, but there are many exceptions. Several of the strongest solar flares ever recorded have happened in the last few years (in a time of minimum sunspot activity) and these resulted in fantastic auroral displays.

WHAT CAUSES THEM?

Aurorae are the result of a long sequence of events. First, activity on the Sun causes the release of a large number of high-speed particles – usually electrons – speeding away at 300–1200 km/sec. If the Earth happens to lie in the direction of this blast, the arriving electrons encounter the magnetic field of the Earth and distort it.

NASA

This image of the Aurora Australis was captured from a space shuttle in May 1991. It is clear from this image that the aurora take place at significant altitudes and that the red aurora are formed at nearly twice the altitude of the green features. We can tell that the auroral features are transparent because much lower cloud layers are visible through the green aurora in the foreground. The tail of the Space Shuttle and the back of its open cargo bay are visible to the right.

Particles racing around in the Earth's magnetic field spiral in around the Earth's magnetic poles and hit the outer reaches of the atmosphere. These collisions cause gas high up in the atmosphere (100–500 km altitude) to produce light. Finally, the observer sees the spectacular result!

Certain locations around the Earth's magnetic poles experience aurora almost every clear night. In the southern hemisphere, these regions are almost completely uninhabited. Of all the locations in Australia, Tasmania is likely to see the aurora most frequently since it is closest to the magnetic pole.

AURORA LEGENDS

Many cultures have developed myths and legends to explain aurorae. The aurorae are often said to be gods, or the souls of departed loved ones.

When an especially active display is visible, it is certainly easy to understand how pre-scientific societies might have associated the random motions and changes of the southern lights with a guiding intelligence.

EARTH–SUN CONNECTION

THE CAUSE of the aurora was unknown until organised observations of the activity on the Sun and regular measurements of the Earth's magnetic field – and the disturbances of it that occur due to the Sun – were made. Richard Carrington made the discovery on September 1, 1859.

Carrington made regular sketches and measurements of the positions of sunspots on the Sun using the projection technique. He was measuring a very large and complex sunspot when he noticed two small patches of intensely bright, white light within the sunspot – this is what we now call a solar flare. Flares are very sudden and brief events lasting no more than a minute or two. Carrington noted that the instruments measuring the strength of the Earth's magnetic field at Kew in England experienced a sudden disturbance at the same time as the flare and that 18 hours later they were again even more dramatically affected and there was a magnificent auroral display. We now know that X-rays arrive at the same time as the light from the flare and that it is the charged particles from the flare which arrive so many hours later.

THE FACTS!

AN AURORAL curtain may be tens or hundreds of kilometres long but less than 100 m thick.

STARS CAN be seen through aurorae, because aurorae are transparent.

WHEN AN auroral curtain appears almost directly overhead, all the rays of the aurora appear to converge to a point.

ACTIVE AURORAE change from second to second, unlike most things in the sky.

THE ARRIVAL of the high-energy particles from a solar flare and the activity produced in the outer reaches of the Earth's atmosphere can actually cause satellite electronics to fail and solar panels on satellites to degrade and produce less power.

THE ATMOSPHERE of the Earth puffs up during periods of high solar activity and this results in satellites experiencing more drag and re-entering the Earth's atmosphere sooner. The International Space Station needs to reboost to higher altitude more frequently at these times.

DURING VERY bright displays, some people report hearing hissing and crackling, but there is no good explanation for how this could occur.

VERY RARE but extremely active Aurora Australis displays can be seen even to the northern boundaries of the Australian continent.

DID YOU KNOW?

THE STARS

THE CONSTELLATIONS
– Patterns of Stars

It is hard to resist "connecting the dots" with stars. Throughout history, people from all cultures have claimed that groups of stars look like animals, people, gods or inanimate objects.

CONSTELLATIONS are handy for navigating, locating where a planet or comet is in the sky, and keeping track of the seasons; however, these star patterns are really only a product of the human imagination.

Over the centuries, all sorts of fanciful constellations were imagined. When astronomers from different countries began communicating their work to each other, it was decided to agree on one common set of constellations. In 1930, a group of 88 constellations was agreed upon and these have been used ever since. Their names are all in Latin, but some of the objects they describe, especially in the southern hemisphere, are more modern. For instance, there is Microscopium (the microscope), Telescopium, (the telescope), and Antlia (the air pump)! Other constellations that you might expect to find include Crux (the cross) and Centaurus (the centaur – a mythological beast that was half man, half horse).

One of the last constellations to be renamed was a huge sailing ship in the southern Milky Way, which used to be called Argo Navis. It took up so much sky that it wasn't very useful for locating objects, so astronomers broke it into four parts: Carina (the keel), Puppis (the stern), Vela (the sheet or sail) and Pyxis (the magnetic compass).

DO CONSTELLATIONS CHANGE IN TIME?

Constellations do change, but it takes tens of thousands of years for most to change in a noticeable way. The stars themselves orbit the centre of our galaxy, but they are also in motion with respect to each other. They may be moving very quickly (over 500,000 km/h in some cases), but most stars are so far away that, even over your entire lifetime, you won't notice a change in position with the unaided eye.

ARE STARS IN A CONSTELLATION RELATED TO EACH OTHER?

The short answer is no. Most of the time, the stars you see in a given constellation are at wildly different distances from each other, are moving through space at different speeds and in different directions, and are different ages. However, in a small number of cases, the stars in a constellation are related to each other. The head of Taurus (the bull), for instance, is mostly comprised of a star cluster in which all of the stars were "born" at the same time and are moving through space together. Most of the stars that make up the figure of Orion (the hunter) are related, as well.

WHAT COLOURS CAN STARS HAVE?

Stars only have a small range of colours. You will never see a violet, green or pink star. The colour of a star appears depends primarily on its surface temperature. The hottest stars (100,000°C) have a slight bluish tinge to their white light. Stars with temperatures of 10,000°C look white. Stars with the surface temperature of the Sun have a slight yellowish tinge to their mostly white light. Cooler stars have progressively more orange and then deeper and deeper red colours. Only the very coolest stars have very obvious visual colour. Modern detectors can see a wider range of the spectrum than the eye can – they can detect objects that are too cool to produce light in the optical part of the spectrum.

WHAT IS THE ZODIAC?

In ancient times, it was thought that the location of the Sun, Moon and planets in the sky affected the lives of people on Earth and the events that unfolded around them. Since the nine planets in our solar system are all orbiting the Sun, they all seem to follow one great circle around the sky. This path intercepts twelve constellations which were studied by the ancients, and these twelve – Aries, Taurus, Gemini, Cancer, Leo, Virgo, Libra, Scorpius, Sagittarius, Capricornus, Aquarius, and Pisces – are collectively known as the Zodiac.

WHY DO CONSTELLATIONS CHANGE THROUGHOUT THE YEAR?

Because the Earth orbits the Sun, at particular times of year the Sun will be in a direction in the sky where it obscures the surrounding constellations from view. Constellations won't be visible due to the sunlight scattered by the Earth's atmosphere. The stars are all still there

This image shows the Milky Way where it passes through the constellations Centaurus, Crux, and Carina. The pink star-forming region contains the massive star eta Carinae.

– they don't go away; we just can't see them. As time goes by, the Earth moves along its orbit and the direction of the Sun changes, preventing a different group of stars from being seen during the daytime. Constellations formerly masked by sunlight then become visible in the pre-dawn sky. Over the course of the year, the Sun appears to move through all of the constellations in the Zodiac as the Earth completes its orbit.

ABORIGINAL CONCEPTIONS OF THE CONSTELLATIONS

Arid conditions during much of the year in some parts of Australia make for wonderfully clear skies, and, in order to avoid the heat and humidity, many Aboriginal peoples travelled and traded at night. Not surprisingly, they developed an intimate familiarity with the night sky and used it to navigate on their journeys.

Some of the constellations used by the Aborigines, the Crux for example, are very similar to those adopted by other cultures. However, Aboriginal people also used the absence of stars, such as the dark obscuring clouds of dust along the Milky Way, or diffuse clouds of starlight, to imagine some creatures they saw in the night sky.

ALPHA CENTAURI

ALPHA CENTAURI is one of the most noticeable stars in the southern sky. It is the closest star to Earth – in fact, it is the closest three stars! It appears to be just one star because they are very far away, 4.3 light-years away from Earth, and the unaided human eye can't separate their light. Two are very bright and orbit each other at 23 times the distance of the Earth from the Sun. The third star is much fainter and was located using a telescope in 1915. It is a very common kind of low-mass star. Despite it being part of the nearest star system, it is still too faint to be seen without a telescope.

THE FACTS!

CONTRARY TO what you might expect, the brightest stars in the night sky are often the most distant.

SOME CONSTELLATIONS can be seen all year round in Australia – others can never be seen.

THERE IS no bright "South Star" equivalent to the northern hemisphere's star Polaris. It is simply a coincidence when any bright star lies close to the position of the south or north point in the sky.

EVEN FROM SPACE, the parts of the sky between stars are not completely dark. Astronauts orbiting outside the Earth's atmosphere would still see the combined light of stars too faint to see individually, as well as sunlight reflected off the dust spread through parts of the solar system by long-disintegrated comets and collisions of asteroids.

STARS IN constellations are far enough away that there is no real change to the patterns as seen from any planet in our solar system. The sky, seen from the nearest star system (alpha Centauri) would still share recognisable constellations – except for Centaurus and the bright star opposite to it in the sky – our Sun.

STARS ARE the perfect way to test your vision or to test the optics in binoculars, telescopes and cameras. A star should always appear to be a point. Any different shape indicates a problem.

CONSTELLATIONS

CRUX –
the Southern Cross

The constellation of Crux is
striking and easily recognised.
Despite its compact size, it
contains three of the brightest
stars in the entire sky.

THE SOUTHERN CROSS lies in the
southernmost reaches of the Milky Way
and is never visible from Europe and the
vast majority of North America. It is truly a
southern splendour!

Crux is highest in the sky at midnight
in May but can be seen throughout
the southern hemisphere winter. The
constellation is close to the position of the
south celestial pole, so it is always above
the horizon for southern Australia.

The southern Milky Way: Carina (top), through Crux and
the Coalsack (centre right) to Centaurus (bottom), taken
from Kangaroo Island. The brightest star, alpha Centauri,
is just to the right of centre at the base of the image.

*Right: This skymap
shows the adopted
borders of the
constellation Crux,
the designations of its
brightest stars, and the
location of The Jewel Box
cluster (here labelled as
NGC 4755). The intensity
of the blue patches
indicates the brightness
of the Milky Way. The
Coalsack, being very
dark, shows up as white
in this skymap. Portions
of adjacent constellations
Centaurus, Circinus,
Musca, and Carina are
also shown.*

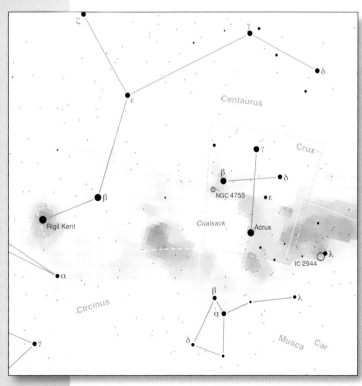

Its visibility at almost any time of night
during the year led to it being used for
navigation by Aborigines and sailors. By
noting the orientation and height of Crux
above the horizon, they were able to
easily deduce which direction was south.
The Milky Way galaxy is a "spiral galaxy",
which means that its young, hot stars are
concentrated in "spiral arms" that converge
at the galaxy's centre. The stars in Crux are
part of the spiral arm in which our Sun (and
solar system) resides. Not all of Crux's four
brightest stars are the same colour or the
same distance away.

The hot, young stars of NGC 4755, The Jewel Box, stand out against the background of Milky Way stars. The orange star is kappa Crucis and is the single star in the cluster that has completed the nuclear fusion of hydrogen in its core.

The three hottest are between 450 and 550 light-years away. Their motions through space are in different directions, which shows that they are unrelated to each other. The third brightest star is a much cooler (but larger) star that is only about 120 light-years away. The three hottest stars are whitish-blue, but the fourth is orange.

THE BOUNDARIES of the constellation of Crux contain the most distinctive example of a "dark nebula" in the entire sky: the Coalsack. As the name suggests, the Coalsack is a region of the sky without stars. While the bright stars can be seen from even the most light-polluted city, spotting the Coalsack requires dark country skies and little or no Moon. It is seen against a starry portion of the Milky Way, so the contrast is strong. We now know that objects like the Coalsack are the result of nearby clouds of opaque dust that block the light from more distant stars.

Aboriginal people form one of their constellations, the head of an Emu, out of the darkness of the Coalsack. The Emu's body follows the dark features projected on the Milky Way to the north, towards the constellation of Scorpius.

With binoculars or a small telescope, it is possible to see within Crux one of the most compact and beautiful clusters of stars in the sky. To the naked eye, they appear to be just one bright star called kappa Crucis, but it also goes by the name of "The Jewel Box". This is one of the youngest star clusters visible in our galaxy, with an estimated age of about 7 million years. It contains roughly 100 stars within a region 20 light-years across. Being an estimated 7,500 light-years away, it is much more distant than the four stars that form Crux. What makes this cluster so stunning is that one of its members is deep orange while the rest are all white or bluish-white. Colour is rarely this obvious in the sky!

THE BRIGHTEST star in Crux, known as Acrux, is a binary star that can be separated in even the smallest telescopes. Both components of this double star are hot and young. They take thousands of years to orbit each other, so no change in their position will be noticed by an observer from year to year. Acrux lies at the base of Crux.

THE SECOND BRIGHTEST star in Crux is also a binary, but the stars orbit so close together that they cannot be separated with any existing telescope. We know the pair exists by examining the spectra of the stars. They orbit each other once in five years and are less distant from one another than the Sun and Saturn.

THE JEWEL BOX cluster was not discovered until 1751–52 by the French astronomer Lacaille. His descriptions of 14 southern constellations made as a result of a trip to the Cape of Good Hope in South Africa have been incorporated in the internationally-recognised list of 88. His constellation list includes "modern" inventions like the Air Pump.

This "star trail" image was defocused in steps so that each star's light would be spread over more image area, improving the eye's ability to distinguish star colours from each other. The stars of Crux are just to the right and above centre. The bright stars alpha and beta Centauri are seen to the left. The colour of alpha Centauri (leftmost) is similar to our Sun. Beta Centauri is much hotter, and therefore bluer.

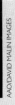

ORION – the Hunter

The most widely recognised constellation in the world must be Orion. In part, this is due to its easily imagined human outline.

AN ALL-TIME FAVOURITE for amateur astronomers and professionals alike, M42, the Great Nebula in Orion, is always pleasing to observe or study. In binoculars or at very low power in a spotting scope, it is clear that the region around the stars of the sword is fuzzy. A little more magnification reveals the swirls of dust and gas as well as the central collection of four bright stars known as the "Trapezium" because of the shape they form.

Observing this region using larger professional telescopes, and at different wavelengths (in the X-ray, ultraviolet, infrared, microwave and radio), and using different kinds of instruments has taught us much of what we know about how stars form.

IT IS also due to Orion's location on the celestial equator, which means that it is visible from both the northern and southern hemispheres. It is best seen from December through to May. In Australia, a common name for the belt and sword of Orion is "The Saucepan".

Unlike the stars in most constellations, the majority of the bright stars in Orion were formed in the same location at the same time and are moving through space in the same direction. The related stars are between

The whole sword of Orion, at the bottom of the constellation, is shown in this image. The naked eye sees three vertical stars, but the camera captures the intricate nebulosity surrounding the middle sword "star" – the "Great Nebula in Orion", M42/43. The nebulosity surrounding the northernmost star in the sword is shown in more detail in the image on the opposite page.

1300 and 1500 light-years away. (The two brightest, Rigel and Betelgeuse, are not part of this collection.)

To astronomers, Orion's most important feature is a thing called the Orion Molecular Cloud. This is a large nearby region of dense gas and dust where there is tremendous star formation activity. New stars are being born as we watch! So large is this region of the sky that it is comparable to the size of the constellation itself.

By studying this cloud in the infrared part

The figure of Orion is outlined by the green lines in this chart. Betelgeuse, in the north-east, and Rigel, in the south-west, are the most prominent stars. The Sword of Orion contains M42, the Great Nebula in Orion. The Horsehead Nebula is located just south of the easternmost Belt star. The outer Milky Way lies along the left boundary of this chart but is much less impressive than the regions near the centre of the galaxy.

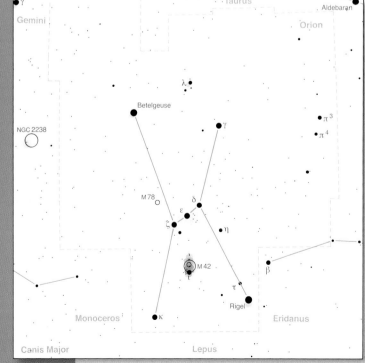

The northernmost star in Orion's sword is surrounded by a delicate combination of emitting gas (pink) and starlight reflected off of dust (blue). This object is known as NGC 1973/75/77. It is about 1500 light-years away from Earth.

THE FACTS!

THE BRIGHTEST STAR in the Trapezium emits so much high-energy ultraviolet radiation that the light strips away the gas from denser parts within the nebula. These objects are seen with material trailing off in the directions opposite the bright star in the Hubble Space Telescope images.

These "cometary tails" – so called because they look like comets – give a three-dimensional feel to images of this region.

NUMEROUS PROTOSTARS – gas and dust clouds on their way to becoming stars – are found in the region around Orion, and disks of dust are also clearly seen in images. These disks will flatten further, condense and fragment into planets at a later stage.

ROUGHLY A THIRD of the stars in this region are known to be binary stars: what looks like a single star to the eye but is actually a close pair (or sometimes even triple star!).

This is what you would see from the country on a dark, moonless night. It matches the chart opposite. The brightest star, Sirius, in the constellation Canis Major, is at lower left and the Milky Way runs diagonally through the image.

of the spectrum (which allows astronomers to see past much of the dust) we have learned a great deal about the process of star formation. The most obvious visual evidence of the cloud is the Great Nebula in Orion, which can be seen through binoculars and small telescopes as a very tight grouping of four bright stars surrounded by a chaotic green, wispy cloud.

The north-east corner of the constellation contains a very interesting star known as Betelgeuse, a red supergiant star. To say that this star is big is an understatement. If placed in the centre of our solar system, the surface of Betelgeuse would be well beyond the orbit of Mars! It is cool by stellar standards, being only 3300°C at the surface. Large telescopes, which do not normally distinguish stars from a point of light, show that Betelgeuse has a

measurable size. Telescopes show that the star has been losing its mass into space. It is certain that Betelgeuse has finished using hydrogen as its source of nuclear energy and is only a short time away from becoming a supernova. Betelgeuse is 430 light-years away, closer than the other stars that form Orion. In the south-west corner of the constellation of Orion is the bright star called Rigel, the seventh-brightest star in the sky. It gives off over 60,000 times more light than our Sun and is about 70 times larger in diameter. For a hot star, this is very large, indicating that Rigel is near the end of its life – it has used up most of the nuclear fuel it has available to produce its energy. In the not-too-distant future, it too will go supernova and blast much of its material back into space. Rigel is about 800 light-years away from the Earth.

ANTARES IS sufficiently close to the ecliptic – the path that the Sun, Moon, and planets follow in the sky – that the Moon passes in front of it periodically. If this occurs in daytime, it is possible to watch the event with a small telescope. At dusk or at night, the event is easy to see with the unaided eye. A series of these events, visible from the southern hemisphere, will occur between 2006 and 2008.

THE GLOBULAR cluster called M4 contains a bizarre pulsar system. A pulsar is a rotating neutron star which is the remnant of a massive star going supernova. This pulsar, known as PSR B1620–26, orbits another stellar remnant – a white dwarf – and orbiting both of these is a planet!

CONSTELLATIONS

SCORPIUS
– the Scorpion

There are only a handful of constellations whose star patterns strongly resemble their imagined creature, and Scorpius is one of them.

THE CONSTELLATION is a large group of bright stars that pass nearly overhead in Australia during the evenings in June through to September, so it is hard to miss!

Scorpius lies very close to the centre of our galaxy, where the combined light of millions of distant stars makes the Milky Way brightest and at its most impressive. The stars of the constellation itself are mostly hot, young stars that reside in the nearest spiral arm to our own.

The most distinctive star in Scorpius is called Antares. Antares is 600 light-years from the Earth. It is the brightest and reddest star in the constellation and is another example of a red supergiant. In fact, the size and surface temperature of Antares are almost identical to that of Orion's Betelgeuse. There is one significant difference: Antares is orbited by a smaller hot star which illuminates the dust that the red supergiant has been throwing off for hundreds of thousands of years.

If you can find Antares with a pair of binoculars, you'll be able to locate one of the nearest and brightest globular clusters, M4. (The M indicates that it was on a list kept by the 18th-century French astronomer Messier, who kept track of objects that could be confused with comets.) Globular clusters are very important in astronomy – they are the largest and oldest-known collections of stars that were born at the same time. There are about 150 such clusters associated with the Milky Way and all are thought to be between 12 and 13 billion years old. (The Sun, by comparison, is slightly under 5 billion years old!) To find M4, locate Antares and then shift the field of your binoculars one or two diameters to the west. You should see a fuzzy patch.

In a small telescope, the fuzz is resolved into hundreds of similar stars packed tightly together – it is quite a sight! The distance of M4 is 7000 light-years, much more distant than Antares.

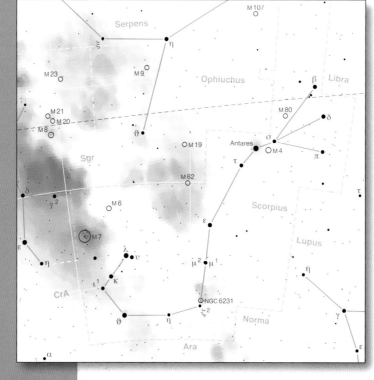

This chart shows the constellation Scorpius and its neighbours. Green lines connect the stars in this constellation in a way similar to a real scorpion shape. The brightest portions of the Milky Way are shown in dark blue. The dust lane obscuring light from the Milky Way runs from middle centre at the bottom to the upper right. Interesting binocular objects are labelled by their NGC or M catalogue numbers.

This region of sky contains objects of many different colours. At the lower left is the red supergiant Antares, surrounded by its own dust and illuminated by its hot, close companion. Directly to the right is the globular cluster M4, a much more massive object containing tens of thousands of old stars. A pinkish emission nebula surrounds sigma Scorpii (σ). (These three objects can be identified in the constellation chart on the opposite page.) The remaining portion of the image lies in the constellation Ophiuchus and contains obscuring dark nebulae and bluish reflection nebulae which occur when light from a nearby star reflects off dust clouds.

THE STAR in the constellation chart labelled "beta" (ß) is a multiple star system. In a small telescope, the brightest two stars are easily separated from each other. The actual distance between them is more than 20 times the diameter of Neptune's orbit around the Sun. The brighter of the two has a companion which can be detected by special techniques but is not visible in a small telescope. Stranger still, both of the visible bright stars are binary stars!

SCORPIUS IS richly rewarding seen through binoculars and small telescopes. It contains a large number of loose open clusters and older and more centrally-concentrated globular clusters. Across the entire sky, only about 100 globular clusters can be seen with small telescopes. Nine of these are found within the adopted borders of Scorpius. Such clusters are most concentrated near the centre of the Milky Way, which lies just to the east of the constellation.

AN EASILY-LOCATED and beautiful star cluster is NGC 6321 which is just north of the double star zeta Scorpii (ζ). It is known as the "Northern Jewel Box" even though Scorpius isn't north of the celestial equator, just much further north than Crux! See the constellation chart.

THE GLOBULAR cluster M4 would be conspicuous to the unaided eye if there weren't a significant amount of obscuring dust along the 7,200 light-year line-of-sight. About 60% of the light is absorbed by dust on its way to us. M4 is one of the very nearest globular clusters to the Sun. Its stars cover an area similar to the size of the Moon's disk in the sky.

A different kind of star cluster, containing a much smaller number of hot, young, bluish-white stars can be found just east of the prominent tail stars in Scorpius. It is known as M7 or "Ptolemy's Cluster", since Ptolemy mentioned it as non-stellar to the unaided eye in the year 130 AD. Ptolemy's Cluster is a pleasure to observe through binoculars, but is paradoxically quite disappointing when seen through a telescope.

THE SPRING
Night Sky

Southern observers are treated to rich portions of the Milky Way that are visible for almost three-quarters of a year.

Spring is the last season in which spectacular sections of the Milky Way are visible. Constellations get lower in the western sky with the passing weeks. For southern observers, the star-poor portions of the sky spell one thing: galaxies! The sky is literally the limit.

High in the sky is the constellation Sculptor. Although it has no stars of remarkable brightness, it contains a cluster of "nearby" galaxies that are easily spotted with small telescopes.

The distances to galaxies, of course, are vastly greater than the distances to stars in our own Milky Way. The "Sculptor Group", at a distance of 10 million light-years, is the nearest collection of galaxies to the group to which our Milky Way belongs. Not surprisingly, the spiral and elliptical galaxies in this group are visually large through a small telescope. Unlike star clusters, the visibility of galaxies is greatly reduced by moonlight and light pollution – a trip to the country is a must, as is a finder chart from a star atlas.

Nearly overhead is the bright star called Fomalhaut. Since we are looking perpendicular to the Milky Way, such a bright star must be nearby – it is only 25 light-years away. In many respects it is like Vega, just a bit cooler (8200 °C) and slightly smaller. Fomalhaut is much brighter in the infrared than would be expected and it is now known to have a large disk of icy dust orbiting it. The presence of a gap in the center of this disk may indicate that planet formation has taken place there. Its proximity and orbiting material make it a prime target for specialised searches for planets by large telescopes in the years ahead.

Almost alone in the south-east is the bright star Achernar. More than 5000 times more luminous than our Sun, it is an extremely rapid rotator. It is the most flattened single star known! The diameter of the star at its equator is twice that from pole to pole! At its equator the gas is moving at 900,000 km/h! It is much larger than the Sun, between 14 and 24 times the size, and emits a great deal of ultraviolet radiation. Achernar is so far out of round that it is extremely challenging to model on a computer – especially given that the motions of the gas inside the star are not known!

Just in case this star is not already complicated enough, it is also known to vary in brightness and to have starspots present on its surface.

Finally, low in the northern sky is the star Deneb. Despite being one of the 20 brightest stars in the sky, it is over 2500 light-years away. Deneb is one of the most impressive stars in our galaxy, emitting as much light as 150,000 Suns. Its temperature is very similar to Vega's but its size is 100 times greater. It has more than 20 times the mass of the Sun and it is likely to become a supernova in a few million years, at most

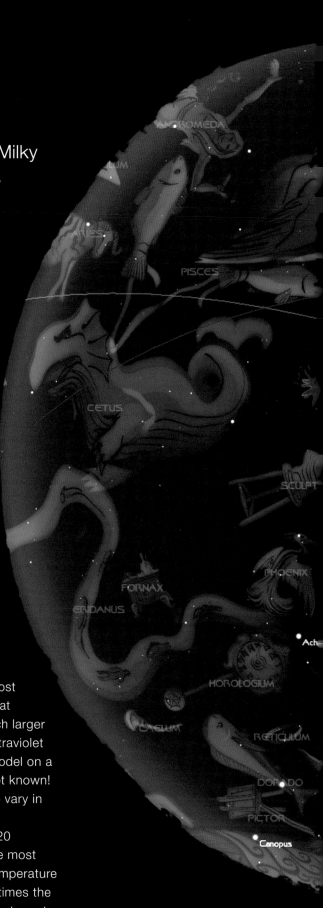

USING THE SKYMAPS!

IN THIS SECTION, the sky at 9 p.m. local (standard) time in Sydney is shown by season.

These are whole skymaps which show the sky as you would see it if you were lying on your back with your head pointing towards the north. East is to the left and south is towards the bottom of the page.

(On maps of the Earth, east would be in the opposite direction but remember that we are now on the surface looking out instead of above it looking down!)

The blue line is the equator. The red line is called the "ecliptic" – it is the path along which the Sun, Moon, and planets appear to travel.

The planets have not been shown because their positions change with time. Consult a website or a monthly astronomy magazine to find out where they are when you are stargazing.

If you are stargazing at midnight, use the skymap for three months later.

A keen observer will note that there is one point on the skymaps whose position remains unchanged – the south celestial pole (not actually marked). It lies in the constellation of Octans and has no bright star nearby. The Earth's axis of rotation points to that position (and to its counterpart in the northern hemisphere).

On all of these maps, the pole is located $^1/_{6th}$ of the way from the southern point on the horizon to the northern point. The whole sky wheels about this point with objects rising in the east and setting in the west. Any stars or objects seen between the pole and the southern horizon remain in the night sky all year.

The constellation art represents the figures imagined based on the arrangement of stars in the constellation. In many cases, they require a great deal of imagination!

THE SUMMER
Night Sky

The bright stars of Orion are immediately recognisable
high in the north-eastern sky.

The bright red supergiant in Orion's shoulder, Betelgeuse, is easy to locate by its distinctive orange hue. If you have a telescope or a pair of binoculars, don't miss an opportunity to find the "Great Nebula in Orion" which is located in Orion's sword. Fuzzy at low magnification (less than 10x), the nebula becomes a fascinating and chaotic patch of wispy gas and dust at higher powers.

The brightest star in the sky is found nearby. Called Sirius, it is only slightly larger and hotter than the Sun and appears bright primarily because it is one of the very closest stars – only 8.6 light-years away. A remarkable aspect of Sirius is that it has an unusual object orbiting it called a "white dwarf". Such an object is known to be the remnant of a star which has exhausted the gas that can be used to maintain its nuclear reactions. Known as "Sirius B", this white dwarf has almost exactly the same mass as the Sun, occupies $1/1,000,000$th of the volume, and gives off only $1/40$th as much light. If Sirius B was still a normal star with its original mass of 3.5 solar masses, it would appear brighter than Venus in the night sky.

Another bright star called Procyon is found very nearby, somewhat lower in the north-eastern sky. Remarkably, Procyon is also very nearby (11.4 light-years) and also has a white dwarf orbiting it!

You may notice that the eastern half of the sky contains a band of brighter stars reaching from the north point on the horizon, through Orion and Sirius and all the way down to the south.

These stars are the most obvious indicators of the Milky Way, the northern half of which is further away from the centre of the galaxy than the distance at which the Sun orbits. This Milky Way portion is not especially impressive because the density of stars decreases as the distance from the galactic centre grows. As you trace the Milky Way towards the south at this time of year, it becomes brighter and more interesting. There are a variety of beautiful and fascinating clusters and nebulae in the constellation Carina, which can be easily spotted with binoculars.

A moonless night at this time of year will allow you to locate the Large and Small Magellanic Clouds – two smaller galaxies orbiting the Milky Way. Some features of these galaxies can be spotted in binoculars, but the Milky Way counterparts of their clusters and nebulae are typically 50 times closer and therefore more impressive. The Large Magellanic Cloud is located between Mensa and Dorado on the skymap. The more compact Small Magellanic Cloud can be found between Hydrus and Tucana.

The western half of the sky, by comparison, seems pretty empty and uninteresting. This is because we are looking in the direction away from the Milky Way. Since there is very little obscuration due to dust in this direction, it is a region of sky where countless distant galaxies can be located with a small telescope. Crux and the two bright, distinctive stars of Centaurus are found very low on the south-eastern horizon.

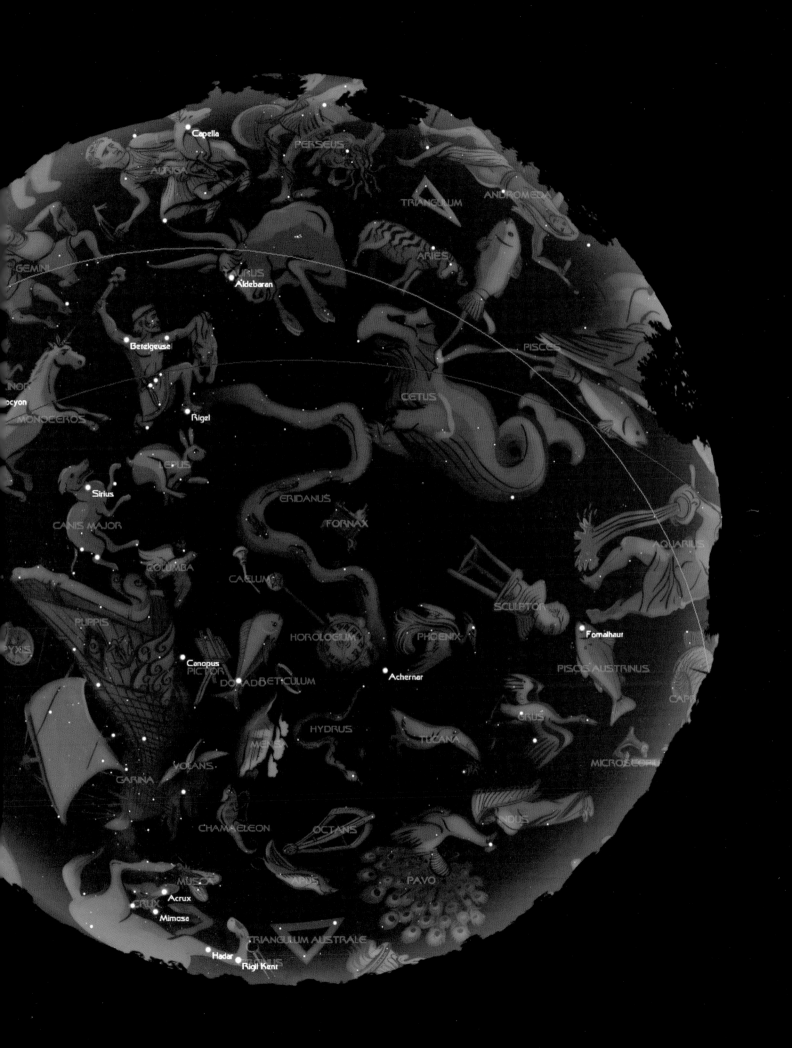

THE AUTUMN
Night Sky

The Milky Way is oriented more east-west at this time of year. Orion and its entourage are seen preparing to set in the west.

The most prominent constellations and objects are again found in the Milky Way. The eye fixes quickly on the two brightest stars in Centaurus: alpha Centauri (also known as Rigil Kent) which is 4.3 light-years away, and beta Centauri (also called Hadar) at 330 light-years. This vividly illustrates how apparent brightness alone tells you very little about the distance of stars! The stars of Crux lie just to the west and of course the large, dark patch of dust known as the Coalsack lies just to the east of the stars of the Southern Cross. The Milky Way seems to be much richer in bright stars along this stretch – and it is.

We are looking at a nearby spiral arm within which there has been the very recent formation of many massive stars. Not too many millions of years from now, this collection of stars will all give rise to supernovae which will rival the Full Moon in total brightness (but all packed into one pinpoint of the sky).

The second brightest star in the sky, Canopus, is high in the south-western sky. As seen from Earth, it shines with only about 40% of the brightness of Sirius but is very different in nature. It actually emits 20,000 times as much light as our Sun and is located 310 light-years away. It is the most luminous star within 700 light-years of us! Part of the reason for it producing so much energy is that it is in a later stage of its nuclear fuel burning and the star's size has increased dramatically as a result of the changes to its core. On this skymap you will notice that any point along the red line of the ecliptic seems to be equally distant from Canopus. Interplanetary space probes make use of Canopus being at a right angle to the orbits of the planets to orient themselves correctly for propulsion and planet encounters.

Two prominent stars north of the Milky Way in the sky are Regulus in Leo and Spica in Virgo. Because both of the stars lie very close to the ecliptic, the Moon and bright planets are sometimes seen near them.

Regulus is the 21st-brightest star in the night sky. It is hotter and larger than the Sun but is still burning hydrogen in its centre – the longest-lived stage of a star's life. Its main claim to fame is that it is spinning so rapidly that it is distinctly not round (although it is not possible to see this with a normal telescope) and the poles of Regulus are significantly more luminous that the equivalent areas near its equator. It is rotating at about 86% of the speed that would tear it apart. For comparison, the Sun's rotation speed at its equator is $1/150$th that of Regulus and all areas on the surface emit equal amounts of light.

Spica is more luminous than Regulus, in fact it is two stars orbiting each other in just over four days, each of which is hotter and more massive than Regulus! (Double stars are quite common.) The stars are orbiting in such a way that they eclipse a portion of each other during each orbit, but the fraction hidden is so small that the light change isn't detectable by the human eye. Spica produces more than 2000 times as much visible light as our Sun but is so hot that it produces ten times that amount in the ultraviolet part of the spectrum.

Arct

LIBRA

Anta

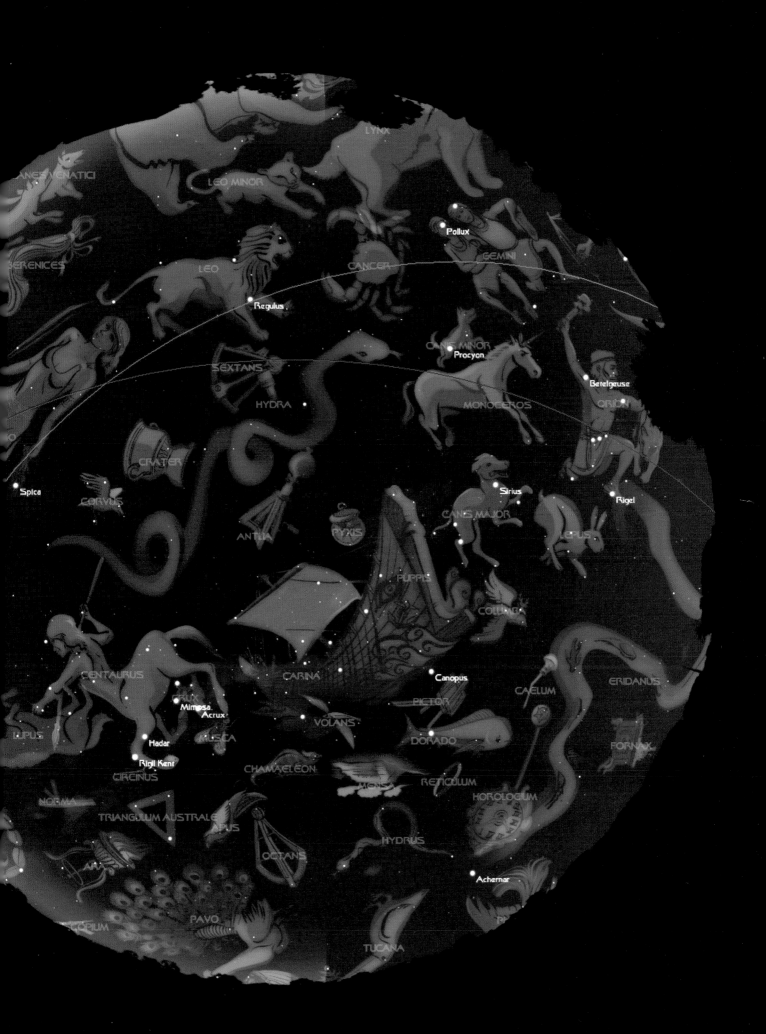

THE WINTER
Night Sky

The southern Milky Way, in all its glory, is the centerpiece of the sky at this time of year.

The bright stars of Centaurus and Crux are still visible, but they have been joined by Scorpius and especially Sagittarius – home to the brightest portions of the Milky Way. Indeed this is the direction of the centre of the Milky Way galaxy and it is no surprise that the stars and dust clouds are dense in this region of the sky. If you are a city-dweller, find a way to get to the country around New Moon at this time of year and you will be treated to a breathtaking sight.

The brightest portions of the Milky Way are sometimes referred to as "star clouds" even though they have no real existence – they are actually just lines of sight with lower amounts of dust obscuration! Still, it is easy to see how they acquired that name. A slow sweep of the Milky Way with a pair of binoculars or a small telescope will reveal dozens of clusters of stars (both of the globular and open varieties) as well as gas clouds and dense patches of dark dust.

The red line representing the ecliptic passes over some of the densest portions of the Milky Way in Sagittarius, meaning that you will sometimes have the pleasure of seeing a planet projected against a rich background of distant stars. With a little patience you may even be able to notice the planet moving from hour to hour against this backdrop. The region between Scorpius's tail and Scutum is especially rewarding for the amateur stargazer.

The centre of the Milky Way galaxy is not visible to the eye or even to large telescopes at visible wavelengths due to dense obscuring dust. It is 28,000 light-years away. It has been possible to

than our Sun, but only about 50% more massive. In total it gives off about 100 times as much light as our Sun. It is relatively nearby at about 37 light-years. As stars go, it is pretty unremarkable!

In the north-east, the very nearby stars Altair and Vega make an appearance. Altair is part of the constellation of Aquila and Vega is in Lyra. Vega has a temperature of 10,000°C and is 25 light-years distant. It is the most luminous of the very nearest stars.

Compared to the Sun, it is three times more massive, 2.5 times the diameter, and produces over 50 times as much light. Altair has a temperature of 7200°C, which makes it hotter than the Sun but cooler than Vega. It is 16.8 light-years away. Like Regulus, it rotates so quickly that it is out of round. Compared to the Sun, Altair actually has about twice the fraction of elements heavier than helium. This is in contrast to most stars which have lower fractions than the Sun.

This is the least favourable time of year to see the Large and Small Magellanic Clouds because they are so close to the horizon. Nonetheless, they remain in the sky year round.

THERE ARE estimated to be tens of billions of stars similar to the Sun in the Milky Way.

THERE ARE roughly 10,000 open clusters in the disk of the Milky Way. Ninety percent of these can't be seen because of the dust in the disk obscuring distant objects.

THERE ARE 150 globular clusters orbiting our Galaxy.

ALL STARS form in clusters out of gas and dust clouds.

Two open clusters in the constellation Gemini are seen in this image. M35 is the large cluster in the upper right. It is young and 2800 light-years away. The redder cluster to the lower left is NGC 2158. It is six times farther away and much older.

THE OUTER REGIONS

INHABITANTS
of the Milky Way

The Milky Way is a fairly typical spiral galaxy. The types of objects found here can also be found in any similar galaxy.

HOWEVER, we are very, very much closer to the inhabitants of the Milky Way so we can see them in much greater detail.

Stars are the most prominent inhabitants of the Milky Way. There are lots of them – literally hundreds of billions. The exact number is unknown since the faintest can only be detected when they are nearby and a large fraction are hidden by dust. Like any other mass, they orbit the large concentration of matter at the centre of the Galaxy. In the disk, the orbits of stars are very close to circular. When stars pass close to each other, their mutual gravitational attraction alters their orbits somewhat. Older stars have had more of these encounters and, on average, can be found higher above the disk as a result.

OPEN CLUSTERS

"Open clusters" are groups of stars which have formed recently in the disk of the Milky Way. They usually do not have more than a thousand solar masses worth of stars in them. In fact, in the act of forming, the cloud usually loses enough mass to prevent the cluster members from staying together.

The stars in an open cluster will gradually scatter away from each other and become part of the general star field population. While they last, the most massive, hottest young stars easily stand out against the background of older stars.

GLOBULAR CLUSTERS

"Globular clusters" are much more impressive groups of stars than open clusters. They typically contain hundreds of thousands to millions of solar masses of stars. Unlike open clusters, they are very old. Most formed at the same time 12–13 billion years ago.

At that time, they would have been fantastically bright, but now the massive stars are long gone. Globular clusters are found concentrated around the centre of the Milky Way, but some have elongated

CANADA-FRANCE-HAWAII TELESCOPE/J.-C. CUILLANDRE/COELUM

The Milky Way has a collection of about 150 globular clusters circling it. This is one of the most interesting called 47 Tucanae because it lies in the constellation of Tucana. It is 15,000 light-years away and contains several million stars.

AAO/DAVID MALIN IMAGES

orbits that take them far above the disk of our Galaxy. About fifty of them can be seen in a small telescope. The very best, richest globular clusters, omega Centauri and 47 Tucanae, can only be seen from the southern hemisphere.

EMISSION NEBULAE

"Emission nebulae" are glowing clouds of gas. They are always found near very young, hot, massive stars and are associated with star-forming regions.

Visually, these nebulae always appear green, although most images show them as pink or red. The eye is very sensitive to green light and the oxygen atoms in nebulae emit this colour of light. Both oxygen and hydrogen also emit red light. Eyes aren't very sensitive to red light, but modern detectors are.

DARK NEBULAE

"Dark nebulae" are regions where there is an apparent absence of stars. In these regions, dust prevents the light from more distant stars from reaching observers on Earth. The dust particles are mainly carbon or silicon and, on average, are only about $1/1000$th of a millimetre across.

The dust clouds are not spread uniformly through space, they seem to mostly be found in long filaments except in star-forming regions.

PLANETARY NEBULAE

"Planetary nebulae" aren't related to planets at all. They are the late stage of the life of a star like the Sun. After such a star has finished using up all of the available nuclear fuel (hydrogen and helium), they become unstable and the outer layers are thrown off into space. The exposed core of the star briefly (for about 50,000 years) illuminates this gas as it expands away into the Milky Way field.

Small planetary nebulae look green (due to oxygen atoms again) and have a small disk, like Uranus and Neptune, hence the name "planetary". The most famous is M57, the "Ring Nebula", in the constellation Lyra.

THE FACTS!

SUPERNOVAE (catastrophic stellar explosions) occur in a galaxy like the Milky Way only every couple of centuries.

THE REMAINS of supernovae, called supernova remnants, can be observed for only about 100,000 years.

SEVERAL DOZEN planetary nebulae in the Milky Way can be found with a small telescope.

THE MILKY WAY is 100,000 light-years across which is one million million million kilometres.

"DARK MATTER" can't be seen. That makes it a rather strange thing to mention. Yet it seems to account for up to 90% of the mass of the Milky Way! We don't know what it is at present – it could be the remnants of massive stars or an as yet undiscovered particle, or a combination of these possibilities. It is one of the mysteries that astronomers are trying to understand. We know that other galaxies have similar amounts of it, whatever it is!

MOST EMISSION NEBULAE change over the course of tens of thousands of years. This is because the pressure inside the glowing part of the nebula is greater than the surroundings. Once the hot stars evolve enough to stop producing ultraviolet light, the gas ceases to glow.

MOST OF THE MATTER that forms a star is stuck with it forever because most stars have less mass than our Sun and do not have a violent end.

Below: The star cluster NGC 346 in the Small Magellanic Cloud surrounded by the dust and glowing gas of its star-forming region, as captured by the Hubble Space Telescope. This is a very young cluster, between 3 and 5 million years old.

STAR BASICS

STELLAR
Nurseries

Stars take so long to use up their nuclear fuel that it seems that they must have been around forever. We know that this isn't true. In fact, we know where they are "born"!

KNOWN AS star formation regions, the parts of galaxies where new stars begin to shine are inevitably dense gas and thick dust. By detecting the infrared light emitted, we can study these compact clouds even before they start to produce energy by fusion. New stars are formed in groups and, for every massive star formed, there are many lower-mass stars created. The gas surrounding the birthplaces of stars is illuminated by the hottest stars and glows with certain characteristic colours.

Gravity is, of course, the force which causes the collapse of gas and dust into new stars, but it is not a simple case of falling into a central place. A gas cloud acts very much like a party balloon. If you try to squeeze it, it doesn't shrink to a smaller size – the pressure and temperature of the air inside increases and it "fights back". When you release your hands, the balloon isn't any smaller, it returns to its original size.

A cloud of pure hydrogen and helium gas with enough material for, say, a few dozen stars wouldn't collapse either. The trick is to have a way for the cloud to lose energy so that it can collapse more easily. Fortunately, nature has figured out a way to do this. Gas, which has heavier elements like oxygen and dust, can convert the energies of motion in the gas particles in the cloud into infrared light, which is then able to escape the star-forming region. As the cloud collapses and gets denser and denser, this process becomes more and more efficient (until just before the star forms!). This gas and dust is so good at getting rid of the excess energy that the centre of the collapsing cloud can be the coldest part, which is the opposite of what you would expect. A disk of material is an inevitable feature of a freshly-formed or forming single star. Astronomers have been able to image these disks directly and they suggest that the later formation of a planetary system around stars is commonplace.

NASA/ESA/A. NOTA (STSCI/ESA)

The best known star-forming region is Orion, at roughly 1500 light-years away. However, there are many, many others. One of the closest is in the constellation Taurus at 450 light-years distant. The southern Milky Way has literally hundreds of stellar nurseries of various sizes, which are revealed by the glowing gas and young, hot stars. Typically these objects are thousands of light-years away.

The most spectacular star-forming region outside the Milky Way is the region called the "Tarantula Nebula" in the Large Magellanic Cloud. Despite being 170,000 light-years away, it is easily visible in binoculars – a tribute to the brightness and number of the stars formed there. When a star-forming region is not very large, the production of massive stars can have the unexpected effect of preventing further star-formation by driving material away with the intense radiation and particle winds of the powerful stars. These stars can also

have the effect of blowing off enough of the remaining gas and dust to prevent the cluster of stars from remaining gravitationally bound to each other. After an orbit of the galaxy, the individual stars in the cluster will have travelled away from each other and their birthplace.

When the very first generation of stars was formed, not long after the Big Bang, it was a different story. These first stars are believed to have been very, very massive, hundreds or thousands of times more massive than our Sun. They were able to cool through a different process that was far less efficient. When they did eventually form, they existed for a very brief interval – probably less than a million years – and produced large quantities of heavier elements, which made collapse much easier for future generations of stars. The gas and dust that collapsed into our Sun and planets had probably been inside a handful of different stars before it formed the Earth.

THE ULTRAVIOLET LIGHT emitted by the most massive stars is so intense that it causes as yet uncollapsed gas and dust clouds to form teardrop shapes pointing back to the massive star.

LIFE IS LESS LIKELY to be found on planets orbiting hot, massive stars because the star does not survive for very long before using up its nuclear fuel. It is expected that higher forms of life require more time to evolve than the few million years that such stars will exist.

ALTHOUGH THEY LOOK substantial in images, the brightest regions of glowing gas in a stellar nursery are far less dense than the best vacuum ever created in a laboratory on Earth.

PLANETARY SYSTEMS have probably not formed around many binary stars because of the lack of stable orbits. The evolution of life likely requires a stable set of atmospheric conditions that a planet "orbiting" in a binary star system would not experience.

IF IT WERE POSSIBLE to see the night sky from inside a stellar nursery, it would look spectacular. There would be a handful of truly brilliant stars, some as bright as the Full Moon, but with all their light concentrated in a point. They would each cast shadows. The dusty dark nebulae would blot out large portions of the sky, maybe more than half, and much of the remainder would be filled with glowing wisps of gas.

Pillars of dense obscuring dust are illuminated by the intense light from the young cluster of stars called M16. Dust, glowing gas, and young, hot stars are frequently seen together. This region is 6,000 light-years distant. The cluster and nebulosity are visible in a small telescope.

CANADA–FRANCE–HAWAII TELESCOPE/J.C. CUILLANDRE/COELUM

THE FACTS!

THE EXISTENCE of atoms which radioactively decay indicates that the Universe cannot have been around forever.

SUPERNOVAE ARE so bright when they explode that for a period of weeks they give off more light than all of the other stars in a galaxy combined.

AS A fusion reactor, the centre of the Sun is very poor. It makes up for its low rate of energy production by the large amount of gas there. Under-achieving as an energy producer allows the Sun to last for 10 billion years. Stars like the sun do not explode when they use up their fuel.

THE SURFACE GRAVITY of a neutron star is 200,000 million times greater than the Earth's. If you weighed 50 kg on Earth, you would weigh 100 million million kg on a neutron star!

THE REMAINS
of Stellar Explosions

Stars don't remain stars forever. The light given off at the surface of a star is due to the energy produced by nuclear reactions in its core. Eventually – and quite dramatically – they run out of gas!

STELLAR NUCLEAR REACTIONS

gradually consume the available fuel, hydrogen gas, turning it into helium. Once the hydrogen is gone, the star attempts to fuse helium into carbon. The most massive stars are able to keep producing energy by fusing heavier and heavier atoms, but not for very much longer. When they attempt to fuse iron, the reaction takes up energy, rather than producing it. In a matter of seconds, the star collapses and produces a supernova and also a bizarre remnant – either a neutron star or a black hole.

The lifetime of the most massive stars is very short, usually just a few million years.

A star like the Sun lasts around ten billion years. Low-mass stars use their fuel up so slowly that they will last for hundreds of billions of year without changing significantly.

Massive stars are responsible for producing all of the elements heavier than carbon. The heaviest atoms are created in a few seconds during a supernova collapse. The resulting explosion spews both heavy elements and unprocessed hydrogen and helium from the outskirts of the star back into the galaxy where it can be incorporated into future generations of stars.

Many radioactive elements are formed in the few seconds during the initial collapse of the star, which later gives rise to the supernova explosion. This instant of heavy, unstable element production allows one to determine how long certain elements have been around since the supernova that created them. In effect, it is a clock that can be used over millions or billions of years.

This process of making heavier elements is a "one-way street". Gradually galaxies become enriched with heavier elements.

Except for hydrogen, all of the atoms in our bodies were created inside earlier generations of stars.

NEUTRON STARS

A neutron star is often created in a supernova explosion. In 1054 AD, Chinese astronomers recorded a "guest star" in what we now call the constellation of Taurus. One thousand years later, we see an intricate, expanding cloud of gas and at the centre is a rotating neutron star – an object which is

A supernova is an enormous explosion that occurs when a massive, older star has exhausted the nuclear fuel in its core, which causes the core to collapse in a spectacular fashion. The supernova remnant Cassiopea A is seen in this false-colour image created from three of NASA's satellite observatories. Visible light appears yellow, infrared light appears red, and X-ray emission appears green and blue. This remnant is 10,000 light-years away and about 325 years old.

NASA/JPL-CALTECH/STScI/CXC/SAO

like an enormous atomic nucleus roughly 15 km in diameter, spinning 30 times per second. It emits pulses of radio waves and optical light that are detected with sensitive modern detectors, giving the name "pulsar" to such neutron stars.

Many such pulsars are known and many more neutron stars, which do not emit pulsed radiation (at least towards the Earth), are expected to exist.

The supernova that occurred in 1987 in the Large Magellanic Cloud confirmed that neutron stars do form during supernovae explosions, because the predicted burst of elusive particles called neutrinos was detected at the time of the event. No pulsar has yet been detected in association with that supernova.

BLACK HOLES

Perhaps the weirdest of all the objects in the Universe is a "black hole".

Once created by supernovae they can grow larger by gaining mass. However, a supernova results in such a cataclysmic explosion of material into space that usually the region is cleared and the black hole is left without anything to "eat"!

A black hole is a region where the gravitational field is so strong that not even light can escape. Since we know that it is impossible to travel faster than the speed of light, this means that nothing can escape!

Such regions are very small. If the Sun were to become a black hole, which it won't, it would be 6 km across instead of its current diameter of 1.4 million km!

Astronomers located the few stellar-sized black holes found so far by noticing how they affect the regions around them. The most direct and conclusive way to find a black hole is when it is one member of a binary star – two stars orbiting their centre of mass. The more massive star of the pair will use up its fuel first and go supernova, creating a black hole. As time goes by, the other star will enter the later stages of its existence and will swell up to become a giant or supergiant star; in the process some of its gas will fall towards the black hole. The gas rarely enters the black hole

directly because it is such a small target. Rather it orbits it in a disk and because of the huge energy it gains falling towards the black hole, it heats up to millions of degrees when it encounters other gas already there. At such temperatures, gas will emit X-rays which can be detected by satellite telescopes.

X-ray light is a sure sign that you have a very compact region containing a lot of energy. The first binary star conclusively shown to have a black hole was called Cygnus X-1 and the visible component is easily seen with binoculars.

Stellar-sized black holes are so small in size that they pose no threat to anything. As far as we know, black holes do not get produced with masses as small as that of our Sun. However, if they did and one were placed at the centre of our solar system, we would orbit it exactly like we orbit the Sun. They wouldn't slurp up the planets or our solar system!

In fact, it is very hard to fall into a black hole, just a slight error in velocity and you would miss it. It is only in the regions very, very close to the boundary of the black hole that more peculiar things happen.

This composite image X-ray and optical image of the region around the pulsar in the "Crab Nebula" reveals the strong influence that the pulsar continues to exert on this supernova remnant. A spectacular timelapse video exists of the month to month changes in the centre of this object.

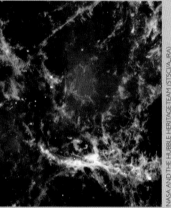

NASA AND THE HUBBLE HERITAGE TEAM (STSCI/AURA)

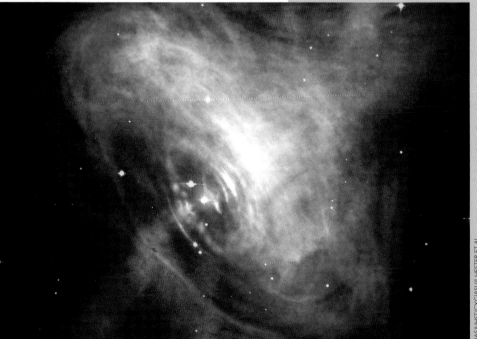

NASA/HST/CXC/ASU/J HESTER ET AL.

THE LARGE Magellanic Cloud extends significantly beyond its brightest regions. It would take 14 Full Moons lined up next to each other to stretch across its diameter in the night sky. For the Small Magellanic Cloud, the corresponding number of Full Moons is eight.

THE FIRST PERSON to methodically study these objects with a telescope was Sir John Herschel at the Cape of Good Hope, South Africa, between 1834 and 1838.

THE MAGELLANIC CLOUDS are associated with an old man and woman by Australian Aborigines.

THE BRIGHTEST, hottest, most massive stars live such short lives that the Magellanic Clouds are the only galaxies near enough that such stars still exist by the time the light has arrived at Earth.

THE MAGELLANIC Clouds are so close to the celestial pole in the southern sky that they can be used to find the direction south.

NASA, ESA AND MARTINO ROMANIELLO (EUROPEAN SOUTHERN OBSERVATORY, GERMANY)

This is a close-up view of a section of the "bar" of the Large Magellanic Cloud revealing two young clusters of stars. The most obvious one is centered and contains hundreds of hot stars as well as a number of redder stars which have finished burning the hydrogen gas in their centres. An even younger and much more concentrated cluster is visible in the lower right. Finally, the blue gas is a remnant of recent star formation in the area.

In this wide-angle view, the Small Magellanic Cloud is to the left and the Large Magellanic Cloud is to the right. All stars in the remainder of the field are members of our own Milky Way galaxy and typically less than 1000 light-years away.

OUR GALAXY'S COMPANIONS

LARGE & SMALL
Magellanic Clouds

Australians are lucky to be able to see the two special companion galaxies to our own, known as the Large and Small Magellanic Clouds.

ONE OR BOTH of these galaxies are visible under dark skies every night of the year; they can never be seen from the northern hemisphere. These galaxies are collections of stars much like the Milky Way, but the galaxies themselves are much smaller. As a result they are less organised in shape and are more easily disturbed when they get close to the Milky Way, or each other. That said, the stars and other objects in these companion galaxies are like other galaxies everywhere else in the Universe.

The name Magellan is associated with these galaxies because they first received attention in Europe as a result of the explorer Magellan's voyage in 1519.

Obviously, Magellan was not the first to see them and they would be well known to the Indigenous people of any country in the southern hemisphere.

The Large Magellanic Cloud is 170,000 light-years away and contains 20 billion times the mass of our Sun. This may seem like a lot, but it is only about 5% of the mass of the Milky Way. The Small Magellanic Cloud is even further away at 200,000 light-years and contains less than $1/3$ of the mass of the Large Magellanic Cloud – that is why it appears less impressive to the naked eye.

Despite their small sizes, both the Large and Small Magellanic Clouds are important to our understanding of the Universe.

From our viewpoint, the Milky Way, the stars and other objects we observe in our own Galaxy are spread far apart. It was not easy for humans many years ago to figure out whether a star was bright just because it was nearby, or whether it was far away but putting out much more energy. All of the objects in each Magellanic Cloud share a common distance for practical purposes. If we see two stars of different brightness in the Small Magellanic Cloud, we know that the brighter one must actually be producing more light.

In 1912, Henrietta Leavitt of the Harvard College Observatory was studying different stars of the Small Magellanic Cloud. She noticed that for a certain type of variable star, the brighter examples always took longer to complete their cycle of variation. She correctly reasoned that if we could figure out the true brightness of any Milky Way counterparts of these stars, we could use them to establish the true distances of galaxies containing similar stars. This discovery led to the recognition that

"spiral nebula" were actually galaxies like our Milky Way at great distances. It also made it possible to determine that galaxies were receding from the Milky Way at a speed proportional to the distance from us – an incredible discovery that completely changed our understanding of the Universe.

The Large and Small Magellanic Clouds have also been important to our understanding of how stars change with time as they use up their gas in nuclear reactions. In recent times, the most important event was the 1987 appearance of a supernova in the Large Magellanic Cloud (known as SN 1987A). Detectors on Earth found unusual particles from it, believed to have been given off in fantastic quantities during a supernova explosion: neutrinos. These particles very rarely interact with normal matter. The experiments, which detected the neutrinos from SN 1987A, were all in the northern hemisphere and the particles had to travel through the Earth to get to them!

THE LARGE MAGELLANIC

Cloud is different from the Small Magellanic Cloud in that the latter's stars contain a smaller fraction of the elements heavier than hydrogen and helium. The heavier elements made in supernovae more easily escape from the lower gravity of the Small Magellanic Cloud.

A spectacular image of most of the Large Magellanic Cloud. Older stars are fainter and distributed more evenly due to a higher total number of gravitational encounters with other stars. The bright star formation containing the Tarantula Nebula is seen to the left center of the image. Other pinkish areas are smaller star formation regions or the remnants of supernovae explosions. The brightest small white clumps are young clusters of stars.

N63A is one of several supernova remnants visible in the Large Magellanic Cloud. The dust and gas from this one is still expanding into its surroundings. The explosion is believed to have occurred 2,000–5,000 years ago. The material from the supernova extends tens of light-years from the original site of the supernova and some of the surrounding gas has been heated to millions of degrees Celsius.

The Small Magellanic Cloud is smaller than its counterpart but still has a number of star formation regions. Its shape has been distorted by passing too close to the Large Magellanic Cloud about 200 million years ago. The concentrated cluster of stars near the top of the image is not part of this galaxy. It is the globular cluster NGC 362, which is only 7% of the distance of the Small Magellanic Cloud.

THE NEAREST spiral galaxy like the Milky Way is M31 in the constellation Andromeda. It is 2.2 million light-years away.

THE MOST distant galaxies found are 13 billion light-years away and formed only a few hundred thousand years after the Big Bang.

THE SOUTHERN SKIES contain some of the most impressive examples of interacting galaxies. The most famous is NGC 5128, also known as Centaurus A.

The nearest very rich cluster of galaxies is found in the constellation Virgo. M87 is a giant elliptical galaxy near the centre of the cluster. It is 60 million light-years away and has 14,000 clusters travelling around it. The Milky Way has less than 200. Bright "spiky" objects are foreground stars.

The galaxy NGC 5128 is 13 million light-years away. A dark dust band that runs across its centre is probably due to a collision with another galaxy.

THE UNIVERSE

GALAXIES
Like & Unlike our Own

The name "galaxy" is attached to any large collection of stars, normally at least 10 million solar masses. Within that broad definition, galaxies can have a variety of shapes, brightnesses, and colours.

THE LARGEST GALAXIES can be 100 times the mass of the Milky Way and as large as six million light-years across.

These large structures are usually found as the centres of rich clusters of galaxies where they have grown by consuming many smaller galaxies over time.

Galaxies are the most distant recognisable objects we can find in the Universe. When the Hubble Space Telescope stared at an apparently blank patch of sky for ten continuous 24-hour periods, it found only a handful of faint, distant stars (all part of the Milky Way) but thousands of distant galaxies, each containing millions and millions of stars! Some of these galaxies are over 12 billion light-years away. The light left these galaxies long before our solar system even formed!

Galaxies form from enormous clouds of gas. When the density of this collapsing gas is high enough, stars can form within it. Astronomers name galaxies depending on their shape. Two very common forms of large galaxies are called elliptical (oval-shaped) and spiral.

This "barred spiral" galaxy is found in the southern constellation of Eridanus (The River) and is called NGC 1300. Many spiral galaxies, including the Milky Way, have bars which form near the centres. NGC 1300 is in a cluster of galaxies which is relatively close at only 60 million light-years away.

NGC 1532, the large spiral galaxy, and NGC 1531, are seen together in this image. The galaxies are located in the constellation Eridanus and are about 55 million light-years away. Obscuring dust and glowing gas (pink) are telltale signs of recent star formation.

ELLIPTICAL GALAXIES

Elliptical galaxies contain little or no remaining gas and stars over 10 billion years old. The brightest stars of this age are no longer burning hydrogen in their cores and appear large and red. These galaxies have smooth circular or oblong outlines.

SPIRAL GALAXIES

Spiral galaxies (like the Milky Way) have large, flat disks of gas and dust that still form stars (although the low-mass stars formed earlier are still there in large numbers). They are the brightest of the galaxies. Newly formed high-mass stars are very bright, whitish-blue and are located in the spiral arms. Consequently, spiral galaxies appear to be bluer than elliptical galaxies. An even larger variety of galaxy appearances are made possible when galaxies collide with each other. These can be spiral-spiral, spiral-elliptical, or elliptical-elliptical collisions. In each case, the result looks different from the others and from the originals! Amusingly, the stars themselves occupy such a fantastically small fraction in the space that they never hit each other when their galaxies collide. The gas, however, is strongly affected. In fact, if one or both of the objects is a spiral galaxy, there is almost always an enormous burst of star formation. Gradually, over the aeons, gas is locked up in stars and removed from the galaxy.

Two spiral galaxies, NGC 2207 and IC 2163, in the constellation Canis Major are seen changing each other during a close encounter. This pair of galaxies is about 115 million light-years away from Earth. The galaxies themselves are roughly 100,000 light-years from each other.

The mass of a galaxy cluster called Abell 2218 distorts the images of a much more distant galaxy, an effect predicted by Einstein. The members of Abell 2218, 2 billion light-years distant, are mostly the yellowish elliptical galaxies. The thin blue and pinkish arcs are the gravitationally-distorted images of background galaxies, between five to ten times more distant.

OUTER SPACE

THE UNIVERSE
Beyond the Milky Way

For the Universe as a whole, what you get is mostly what you *don't* see! It is now known that a great deal of the matter and energy in the Universe is tied up in objects that do not emit light of any kind.

DARK MATTER

By taking spectra of stars or portions of other galaxies, it is easy to determine how quickly they are moving towards you or away from Earth. To determine how much mass must be inside the orbit of a star, we combine the information we have on motion with the distances from the centre of the Galaxy in question. When this mass is compared with the mass of all of the stars, dust, and gas that can be detected, there is a huge discrepancy. There is more mass present than can be seen in any kind of light! While this difference is sometimes called the problem of the "missing mass", it is really a problem of missing light! How much is missing? Well over 90%!

We are aware of some things that contain a lot of mass but give off very little, or no, light. They are white dwarfs, neutron stars, and black holes. These remnants of stars do qualify as "dark matter" but the problem is that if they were created in the numbers required to explain the missing mass, there would be all sorts of additional effects which could be easily seen – like having a far greater abundance of heavy elements.

There is an extremely difficult to detect particle called a neutrino, which is produced during supernovae explosions and radioactive decays. Unfortunately, it is now known from measurement that they can't be the answer either!

So far astrophysicists have only eliminated possibilities. The culprit (or culprits!) have not yet been found. At the present time, the leading idea is that a very massive particle exists, not yet directly detected, which interacts very little with ordinary matter like protons, neutrons, and electrons. Time will tell if we are on the right track.

DARK ENERGY

Not long ago, astrophysicists thought that "dark matter" was the key remaining problem to understanding how the Universe changed with time. However, several different kinds of observations now suggest an even greater influence for the Universe in the long term is a property that causes it to expand more and more rapidly with time.

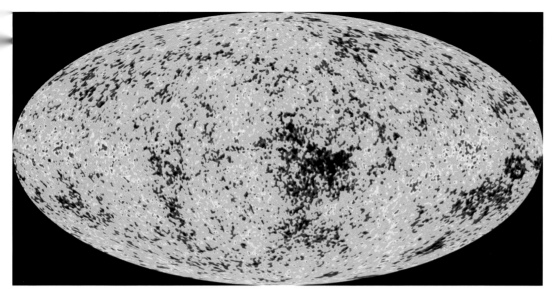

This is an image from NASA's WMAP satellite of the Universe at a very young age – 13.5 billion years ago. This is the most distant observation currently possible and it shows the temperature differences all around the sky roughly 200,000 years after the Big Bang. The redder regions are denser and will collapse to form galaxy clusters.

NEW AND better telescopes, instruments, and satellites inevitably lead to new discoveries and, ultimately, a better understanding of our Universe.

THE VARIATIONS in the microwave light from cosmic background radiation are less than $1/100$th of 1%.

THE EXPANSION of the Universe began 13.7 billion years ago with a reported margin of error of only 1%!

THE UNIVERSE appears to be 4% regular matter, 23% dark matter and 73% dark energy.

THE FIRST stars formed 200 million years after the Big Bang.

This is exactly the opposite of what you might expect in what appears to the eye to be a pretty empty Universe. If you throw a ball into the air, you will see that the effect of the Earth's gravity is to cause the ball to slow down in its climb and reverse direction to fall back to the ground.

The presence of matter and energy in the Universe was expected to have a similar effect in reducing how quickly it expands as time goes by. This latest evidence suggests just the opposite!

What is this "dark energy", which acts like repulsive gravity when the Universe is large enough? No one knows. It is one of the most interesting problems in astrophysics at the present time. Einstein's theory of general relativity does allow such a property of the Universe but doesn't give any clue as to how strong it might be or whether it would repel or attract, so that isn't much help.

COSMIC BACKGROUND RADIATION

We know the Universe is expanding and we have observed and expect that it was hotter and denser at earlier times. If we look back far enough (to a time 400,000 years after the Big Bang) when the Universe was hot enough that electrons could not stay attached to their atomic nuclei, we would see a luminous fog of light, similar in brightness to the disk of the Sun, in all directions.

At that time, and at earlier times, the Universe was not transparent, light would bounce off all of the freed electrons.

This is what is seen, but not in visible light. The expansion of the Universe has caused all of this light to be shifted into the far infrared, a region of the spectrum where microwaves, like the ones in a countertop microwave oven, are found.

The background of radiation for the entire Universe was found by scientists who were developing microwave equipment for telecommunications purposes. They detected a source of noise that wasn't coming from their equipment and seemed to be uniformly strong everywhere in the sky – they also shared the 1978 Nobel Prize in Physics for the discovery!

Unknown to them, other astrophysicists had predicted that a Big Bang model of the Universe would produce such a feature.

Satellites that measure this cosmic radiation far more precisely have now detected the very small variations in its brightness from place to place in the sky. The distribution and size of these variations gives us important information on how the Universe changed with time.

This image contains a cluster of about fifty large galaxies. The two brightest objects are stars in the Milky Way. Clusters of galaxies are common in the sky, and like clusters of stars, gravity is responsible for these structures. The range of brightness and forms is clearly visible.

GLOSSARY

ASTRONOMY The study of the Universe and its contents beyond the bounds of the Earth's atmosphere.

ATMOSPHERE The outermost gaseous layers of a planet, star or natural satellite (moon).

ATOMS The smallest unit of a chemical element.

COMET A partially-vaporised icy body that develops a diffuse envelope of dust, gas and tails when it nears the Sun.

CONSTELLATION Any of the various groups of stars and sky regions, such as Crux, that have been defined by astronomers.

DARK ENERGY Astronomers have yet to figure out what this is, but it acts like repulsive gravity and dominates over long distances.

DARK MATTER Matter which emits little or no light, but is known to exist because of its gravitational pull.

ECLIPSE Where the light from a celestial body is temporarily cut off by the presence of another. A solar eclipse can occur only at New Moon. A lunar eclipse occurs when the Moon passes into the shadow of the Earth.

ELECTROMAGNETIC RADIATION A form of energy that can travel through a vacuum. It consists of linked and rapidly varying electric and magnetic fields.

ELECTRON An elementary particle with negative charge which orbits the nucleus of an atom.

FUSION A thermonuclear reaction process which results in nuclei of light atoms joining to form nuclei of heavier atoms.

GAS Mobile atoms or molecules which move past each other with a speed determined by temperature.

GALAXY A family of stars (and other matter) held together by their mutual gravitational attraction.

GLOBULAR CLUSTER Hundreds of thousands, even millions, of stars that are densely packed together and bound by mutual gravitational attraction.

GRAVITATIONAL ASSIST/SLINGSHOT Using the gravitational field of a planet to change the direction and speed of a spacecraft without consuming any fuel.

GRAVITY The force of attraction that acts between all masses.

HELIUM A very light element first discovered in the Sun's atmosphere. It never forms a solid.

HYDROGEN The lightest element. Its most common form is a proton being orbited by an electron.

IGNEOUS ROCK Rock formed directly by solidification and cooling from a molten state.

LATITUDE Angular distance north or south of the equator in a spherical system.

LIGHT-YEAR The distance that light travels in an Earth year.

LONGITUDE Angular distance east or west from an arbitrary zero point in a spherical coordinate system.

LIQUID A collection of freely-moving atoms or molecules that are weakly bound to each other and maintain a surface.

MAGNETIC FIELD A property produced by moving electrical charges.

MASS The property of a body commonly, but inadequately, defined as the measure of the quantity of matter in it. Associated with resistance to motion change.

MATTER The substance of which physical objects are composed.

METAMORPHIC ROCK Sedimentary or igneous rock that has been deeply buried and changed by heat and/or pressure.

METEOR A particle of dust or rock from space that enters the Earth's upper atmosphere displaying a brief luminous trail, also known as a "shooting star".

METEORITE The recovered fragment of a meteor that has survived passing through the Earth's atmosphere.

MOLECULE The smallest collection of bound atoms which has certain chemical properties.

NEBULA A cloud of interstellar dust and/or gas.

NEUTRINO An elementary particle with both small mass and zero electric charge.

NEUTRON An uncharged massive particle present in the nuclei of all atoms except for the most common form of hydrogen.

NEUTRON STAR A star that has collapsed under gravity and consists almost entirely of neutrons.

NUCLEAR REACTIONS Any reaction that involves a change in the structure or energy state of the nucleus of an atom (which contain protons and neutrons).

NUCLEUS A central thing or part about which other parts or things are grouped. Sometimes used to name the most condensed portion of the head of a comet.

ORBIT The path that a moving body takes in a gravitational field.

PLANET A massive astronomical body orbiting the Sun.

POLES The two points where the extended axis of the Earth pierces the celestial sphere, about which the stars appear to revolve.

PROTON A massive, positively-charged particle present in every atomic nucleus.

RADIATION The process of emission of matter or energy. For matter, emission and multiplying of particles or photons by a radioactive substance.

RED SUPERGIANT A star that has evolved and expanded in size. The resulting change in surface temperature makes it appear red.

RING SYSTEM Ring structures, composed of numerous individually orbiting particles of dust or ice, present around the four largest outer planets – Saturn, Uranus, Jupiter and Neptune.

SATELLITE Any body orbiting around a larger parent body.

SEDIMENTARY ROCK Rock made of fragments of pre-existing rock deposited as a sheet.

SOLID A collection of atoms or molecules which are bound to each other strongly enough to preserve form and prevent relative motion.

SPECTRUM OF A STAR The distribution of colours in starlight. Detailed examination of these colours reveals a great deal about stellar physical conditions.

SUPERNOVA A star that has collapsed or exploded causing an enormous increase in brightness and ejection of matter and energy.

UNIVERSE The collection of all matter and radiation which could possible influence a system.

WHITE DWARF A body that is created when a lower-mass star exhausts its sources of fuel for thermonuclear fusion.